The Iconography of Sir Isaac Newton to 1800

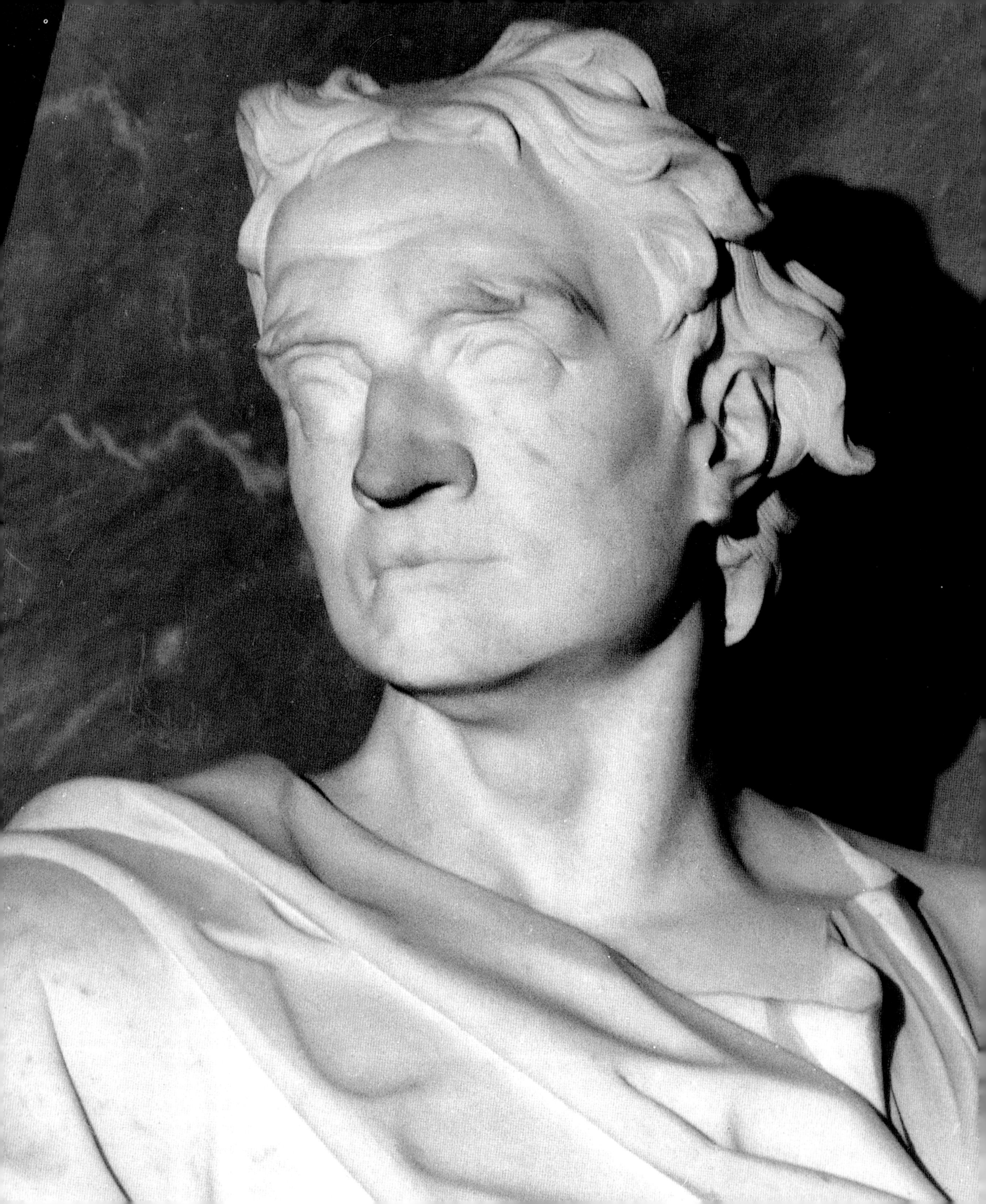

The Iconography of Sir Isaac Newton
to 1800

Milo Keynes

THE BOYDELL PRESS
in association with Trinity College, Cambridge

The author acknowledges the generous support of Trinity College, Cambridge, and a Keynes Family Trust in the publication of this book

First published 2005
The Boydell Press, Woodbridge in association with Trinity College, Cambridge

ISBN 1 84383 133 3

The Boydell Press is an imprint of Boydell & Brewer Ltd
PO Box 9, Woodbridge, Suffolk IP12 3DF, UK
and of Boydell & Brewer Inc.
668 Mt. Hope Avenue, Rochester, NY 14620, USA
website: www.boydellandbrewer.com

A CiP catalogue record for this book is available from the British Library

Library of Congress Cataloging-in-Publication Data

Keynes, W. Milo (William Milo)
The iconography of Sir Isaac Newton to 1800 / Milo Keynes.
p. cm.
Includes bibliographical references and index.
Summary: "Catalogue and iconography of the extraordinary wealth of images of Sir Isaac Newton, both before and after his death" – Provided by publisher.
ISBN 1-84383-133-3 (hardback : alk. paper)
1. Newton, Isaac, Sir, 1642-1727 – Portraits – Catalogs. I. Title.

N7628.N48A4 2005
530'.092 – dc22

2004023329

Designed by Jim Reader
Design and production in association with Book Production Consultants plc,
25–27 High Street, Chesterton, Cambridge, CB4 1ND
www.bpccam.co.uk

Printed and bound by Kyodo Printing Co (Singapore) Pte Ltd.

FRONTISPIECE: The head of Newton from the marble monument in Westminster Abbey carved by Rysbrack, 1731 **D.5a**.

CONTENTS

COLOUR PLATES

ACKNOWLEDGEMENTS

The first acknowledgement must go to **Trinity College, Cambridge**, where Isaac Newton was a member for forty-two years, for encouragement and support in the compiling of this iconography, and for the grant with which to illustrate it. The librarian, Dr David McKitterick, and the staff of the Wren Library, notably Joanna Ball, John Marais, Jonathan Smith and, in the first two years, Alison Sproston, have provided indispensable help in dealing with the intricate problems so constantly raised in such detective work. Thanks are also due to Sir Michael Atiyah, FRS; Professor Dennis Green; Dr Neil Hopkinson; Sir Andrew Huxley, FRS; Professor Simon Keynes; the late Dr Peter Laslett and Paul Simm; and to Les Goodey for the photographic work.

This iconography would have been impossible without the very necessary and freely given help of many experts. It has involved the History of Science as well as the History of Art, and for the former I have had the constant help of Professor Michael Hunter and Scott Mandelbrote, who, as well as Dr Elisabeth Leedham-Green, have also kindly read the text, suggesting improvements and pointing out infelicities.

The icons have come from a wide variety of sources, but mainly from galleries, museums and institutions. They have been classified into nine sections, each requiring its own expert appraisal. Those who have been greatly involved with this can be especially thanked. They include Charles Avery, Malcolm Baker, Paul Cox, Mordechai Feingold, Geoffrey Fisher, Alastair Laing, Christopher Lloyd, Robin Reilly, Duncan Robinson, Malcolm Rogers, William Schupbach, David Scrase and J. Douglas Stewart.

An advantage of working in Cambridge, besides the **University**, **Trinity College** and **Whipple libraries**, has been the **Fitzwilliam Museum** where, despite its partial closure, I have had the valuable help of Dr Mark Blackburn, Elizabeth Fielden, Jane Munro, Dr Julia Poole and David Scrase. Other institutions have been most helpful and can also be thanked – **American Philosophical Society**: Sue Anne Prince, Curator; **Aston Hall, Birmingham**: Chris Rice, Curator; **British Museum**: Aileen Dawson, Curator, Department of Medieval and Later Antiquities; **Christie's**: Francis Russell; **Courtauld Institute of Art**: John Sunderland, Witt Library; Geoffrey Fisher, Conway Library; **English Heritage**: Julius Bryant; **Institute of Astronomy, Cambridge**: Dr D. W. Dewhirst; **King's College, Cambridge**: Peter Jones, Librarian; **Louth Naturalists', Antiquarian and Literary Society**: David N. Robinson;

Magdalene College, Cambridge: Dr Richard Luckett, Pepys Librarian; **National Portrait Gallery**: Dr Catherine MacLeod; Paul Cox, Heinz Archive; **National Trust**: Tim Knox, Architectural Historian; Alastair Laing, Adviser on Pictures and Sculptures; **Royal Astronomical Society**: P. D. Hingley, Librarian; **The Royal Collection**: Christopher Lloyd, Surveyor of the Queen's Pictures; Jonathan Marsden, Deputy Surveyor of the Queen's Works of Art; **Royal Institution**: Dr F. A. J. L. James; **Royal Mint**: G. P. Dyer, Curator; **Royal Society**: Joanna Corden, Archivist; **Salisbury and South Wiltshire Museum**: P. R. Saunders, Director; **Sotheby's**: James Miller; Julien Stock; Simon Stock; **Spalding Gentleman's Society**: J. W. Belsham; **Temple Newsam House, Leeds Museums and Galleries**: Adam White, Curator; **Uffizi Gallery, Florence**: Dr Marco Chiarini; **Victoria and Albert Museum**: Malcolm Baker; R. J. C. Hildyard, Department of Ceramics and Glass; **Wedgwood Museum**: Gaye Blake Roberts, Curator; **Wellcome Library**: William Schupbach, Iconographic Collections; **Westminster Abbey**: Dr Tony Trowles, Librarian; **Yale Center for British Art**: Tim Goodhue, Collections Manager; **York City Art Gallery**: Richard Green, Curator.

I am also grateful to the owners and many others who by their help have allowed this iconography to be, it is hoped, an accurate record. They include Dame Gillian Beer; Professor David Bindman; Lord Balfour of Burleigh; Lizzie Bromley-Davenport; Johnny Bute; Justin Coldwell; Frank Collieson; the late Sir Alan Cook, FRS; Dr Nicholas Cronk; Roy Davids; Professor Eamon Duffy; Lord Egremont; Professor H. Erskine-Hill; Katharine Eustace; Richard Farleigh; Sir John Guinness; Professor A. Rupert Hall; Richard Hicks; Christopher Hogwood; Peter Jones; John Kenworthy-Browne; Dr Desmond King-Hele, FRS; G. Knight Boyer; Cecily Littleton; Charles Littleton; Sir Geoffrey Lloyd; Sandra, Countess of Macclesfield; Earl of Mansfield; James Methuen-Campbell; James Mulraine; Neill Petersen; Earl of Portsmouth; Dr Ingrid Roscoe; Emma Rothschild; Lord Rothschild; Hon. Georgina Stonor; Lord Thomson of Fleet; Dr P. J. Thwaites; Dr V. M. Welter and Camilla West.

Finally, acknowledgement must be made to the incidence of non-scholarly serendipity in the solving of some of the problems raised in this iconography. This particularly occurred in the unravelling of the portrait of Newton in the Uffizi Gallery (**III-7**).

INTRODUCTION

There are numerous accounts of the works produced by individual artists, but few devoted to the portraiture of a single person. In most instances there are not enough different paintings, engravings and sculptures available for such a study, with a scant number of images done from life on which reliably to base other icons. The importance of historical portraits can hardly be exaggerated, and usually our knowledge of how a person really looked is limited and likely to have come from a small number of authentic sources.

To quote from R. E. W. Maddison (1901–1993) in his careful study of the portraiture of Robert Boyle, FRS (1627–1691):

> posterity has an obligation to its famous men, not only to understand the conditions in which they flourished, but also to preserve the more personal details concerning their lives. One inevitable consequence of making a prolonged study of a person's life and work is the growth of a desire to know more of his personality as well as something about his appearance.[1]

For Samuel Johnson (1709–1784) the value of (contemporary) portraiture lay

> in diffusing friendship, in reviving tenderness, in quickening the affection of the absent, and continuing the presence of the dead.[2]

A good portrait may tell us more than a long description, and not a few feel that they appreciate better the work of a man when they have his image before them. There is no reliable representation of any of the medieval heroes of science. From the sixteenth century onward the art of the woodcut and later of the copperplate induced men to attach portraits of themselves to their printed books. Even when paintings are lacking, we thus have adequate portraits of most of the moderns, with the notable exception of Robert Hooke, FRS (1635-1703).

There has been no previous thorough iconography of Sir Isaac Newton (25 December 1642 – 20 March 1727). The text of this book, which involves both the History of Science and the History of Art, includes many biographical details. In his inaugural lecture in 1980 to the University of Cambridge, Professor Derek Beales pointed out the importance of biography to the study and writing of history.[3] In the same way historical portraits may yield significant biographical information, leading here to worthwhile, if sometimes discursive, comment. The chapter by D. E. Smith on 'Portraits of Sir Isaac Newton' in the memorial volume of the Mathematical Association in 1927 was only concerned with the engraved portraits.[4] Written in 1941 for the celebration of the tercentenary of Newton's birth, which had to be abandoned because of World War II, the paper by F. E. Brasch on 'Newton's portraits and statues' is brief, with only four and a half pages of text and illustrated by only five of the authentic portraits.[5]

It is uncommon to have so many *ad vivum* portraits as are represented in this iconography, admittedly all done, with one exception, after Newton had reached the age of fifty-nine. It demonstrates the importance that he attached, either by donating his own portraits or by agreeing to sittings, to the commissioning of appropriate images of himself, while also providing copies for his family. It emphasises his immense fame, and over the years the increasingly great demand for the display of his image, at first mainly by his friends and acquaintances, but then far more widely after his death. It shows the extensive multiplication of portrait replicas, not necessarily by the original artists or from their studios, and of engravings made after oil portraits in the eighteenth century. It was in the early part of that century, too, that portrait busts began to be displayed independently of, and no longer made predominantly for, church monuments.

The icons done from life

Although Newton lived for fifty-seven years in the seventeenth century, only one drawing and one painting, besides four copies of it, were made of him in that century (**I-1**, **II-1**–**II-5**). The drawing was done by David Loggan (1635–1700) *c.*1677 but has long been missing, and the portrait in oil was painted in 1689 by Sir Godfrey Kneller (1646–1723) when Newton was forty-six years old.

After thirty-five years at Trinity College, Cambridge, Newton moved to London in 1696 to become Warden of the Mint, before being raised to the position of Master of the Mint in 1699, and President of the Royal Society in 1703. In 1702, after a gap of thirteen years, Kneller painted him again when he was aged fifty-nine (**III-1**), a third time in 1719 (**VIII-1**), and a fourth time in 1720 (**IX-1**). As well, his portrait was done ten more times by eight other artists between 1709 and his death in 1727 at the age of eighty-four. This gives a total of fourteen paintings of him during his life time. Between 1717 and 1726, William Stukeley, MD, FRS (1687–1765) made several drawings, of which only three appear to have survived (**XI-1**, **XI-2** and **XI-3**).

Besides the *ad vivum* paintings and drawings and seven copies of the death mask (**C.1**–**C.7**), there are two busts, as well as two reliefs, carved in ivory by David le Marchand (1674–1726). One bust was done in 1714 (**A.1**) and the other in 1718 (**A.2**), with the two reliefs earlier between 1702 and1710 (**B.1** and **B.3**). From inscriptions on them, we know that the two busts and the relief **B.1** were carved from life, with Newton paying one hundred guineas to le Marchand for **A.2** , and keeping it in his possession until he died. There is another ivory bust, though not certainly made by le Marchand, dated 1720–5 (**A.4**).

The iconography

The iconography is divided into two parts, **Portraits** and **Sculptures**, in which the *ad vivum* icons are separated from the posthumous ones traced up to 1800. In Part I, there are autograph versions of four of the fourteen portraits in oil done in life (**III-1**, **IV-2**, **XIII-2** and **XIV-2**), and preliminary studies for two (**XIII-1** and **XIV-1**). Eleven copies after six of the portraits were made in the studios of the original artists, twenty-two copies (often of inferior quality and done by jobbing copyists) after seven by other artists, and thirty-one engravings after six of them. Four of the engravings were made during Newton's life, two in 1712 from portraits **III-1** and **IV-2**, and two in 1726 from **XIII-2**, of which one was the frontispiece by George Vertue (1684–1756) for the third edition of *Principia Mathematica*, 1726 (**XLVIII-1**).

The section on posthumous portraits numbers nineteen, of which two are in allegorical scenes. Besides the thirty-one engravings after the *ad vivum* portraits, there are fourteen other engravings, five of them allegorical. Not including ten spurious and doubtful portraits (**XXXII** to **XLII**), this gives a total of 122 icons up to 1800 catalogued in Part I.

In Part II, the posthumous sculptures greatly predominate over the five *ad vivum* ones already mentioned. In section **A**, besides the three ivory busts by or after le Marchand, there are thirty-four portrait busts in terracotta, marble, plaster of Paris, bronze or ivory, giving a total of forty-one icons when four paintings are included. In section **B**, there is a total of twenty-three reliefs in ivory, plaster of Paris, wood, wax, terracotta or fired clay, and in section **C**, the seven death masks. There are thirteen ceramics, mainly Wedgwood busts and portrait medallions, grouped with four waxes in section **E**, eight cameos and intaglios in **F**, eight portrait medals and tokens in **G**, with the monument in Westminster Abbey, two statues, two allegorical paintings, one allegorical fresco, and design of a cenotaph placed in section **D**. This adds up to a further 109 eighteenth-century icons in Part II, giving with Part I a total of 231 icons of Isaac Newton up to 1800.

Comparison with the iconographies of Harvey, Boyle, Pope and Garrick

There have been few such catalogues in the past, but Newton's iconography can be compared with those of two seventeenth-century figures, the physician William Harvey, MD (1578–1657) and the natural philosopher Robert Boyle, as well as with those of the poet Alexander Pope (1688–1744) and the actor-manager David Garrick (1716–1779).[1,6,7,8]

The earliest portrait of Harvey, *c.*1620 when he was aged about forty-three, is by Daniel Mytens (1590–1642), in the National Portrait Gallery. Another *ad vivum* portrait, probably done in 1651 when Harvey was seventy-three, and for which the drawing is missing, is a copperplate etching, doubtfully attributed to Richard Gaywood (*fl.*1650–1680), intended for insertion in the first edition of *De Generatione Animalium* (1651). There is a seventeenth-century copy in oil after it in the Kent and Canterbury Hospital and another one in the National Portrait Gallery.[6]

An *ad vivum* portrait in oil, arguably attributed to Cornelius Janssen (1593–1664), has been in the possession of the Royal College of Physicians since before 1666: the engraved frontispiece for *Anatomical Exercitations* (1653) was made after it by William Faithorne, the elder (1616–1691), in 1653, and the same year Richard Gaywood used the image in another engraving, besides copies in oil in 1735 and 1773 and a frontispiece engraving by John Hall (1737–1797) for the *Opera Omnia* in 1766.

A third portrait in oil of Harvey, thought likely to have been painted by Wilhelm van Bemmel (1630–1708) in Rome *c.*1650, is in the Hunterian Collection in Glasgow. This was engraved by Jacob Houbracken (1698–1780) in 1738 when it belonged to Dr Richard Mead, MD, FRS (1673–1754); the engraving was subsequently used to illustrate *The Heads of Illustrious Persons of Great Britain* ... (1743–52) by Thomas Birch, FRS (1705–1766).[9] At the sale of Mead's collection in March 1754, the painting was bought for William Hunter, MD, FRS (1718–1783) by the dealer, Mr Oram, who at the same sale also purchased for him a copy of Kneller's second portrait of Newton (**III-2**).[10] There are four oil copies and four more engravings after the Hunterian portrait, and also a marble bust in the Royal College of Physicians carved after it in 1739 by Peter Scheemakers (1691–1771) with three replicas.

An unattributed marble bust of Harvey was destroyed in the Royal College of Physicians in the Great Fire of 1666. Edward Marshall (1598–1674) carved the bust placed in Harvey's monument in Hempstead Church, Essex, in 1657, which was most likely taken from the portrait in the Royal College of Physicians.[6] In addition, six seventeenth- and eighteenth-century portraits in oil have been traced, and four other eighteenth-century engravings identified.

Harvey died at the age of seventy-nine when Newton was aged fifteen and Robert Boyle thirty years old. In summary, there is a total of thirty-six icons of him up to 1800, consisting of seventeen portraits in oil, thirteen prints and six busts, including four *ad vivum* images.

Robert Boyle was fifteen years older than Newton, and aged sixty-four when he died thirty-six years earlier than he in 1691. His first portrait, an *ad vivum* drawing on vellum in pen, pencil and wash at the age of thirty-seven in 1664, was by Faithorne, but strangely his engraving from it is not in reverse. A re-engraving (again not in reverse) was made by François Diodati (1647–1690) for a frontispiece to Boyle's *Opera Varia* in 1680, and from this a further re-engraving was done by Isobella Piccini (*fl.*1665–1692) for a frontispiece to Boyle's *Opera Omnia* in 1696–97. There is a portrait in oil by an unknown artist after the Diodati engraving in the Uffizi Gallery, and of similar size to the portrait of Newton there – see **III-7**.

There are several copies of a well-authenticated oil portrait of Boyle by Johann Kerseboom (*fl.*1683–1708) done from a sitting in 1689,[11] and there is also a portrait by John Riley (1646–1691) in the Royal Society. Jean Cavalier (*fl.*1680–1707) made a bust-length ivory carving in 1690, now lost, but recorded in a medal cast by Carl Berch (1707–1777) in 1729. That gives four *ad vivum* portraits, a similar number to the four of Harvey, as three other sittings to Sir Peter Lely (1618–1680), Gerard Soest (d.1681) and Mary Beale (1632–1697) are not certain.

A portrait-bust was carved in marble by G. B. Guelfi (1690 to after 1734) in 1732, and there is another one, by 1749, in the Library at Trinity College, Dublin, now attributed to a collaboration between Peter Scheemakers and L. F. Roubiliac (?1705–1762).[12] These with twelve more oil paintings, five other sculptures and nineteen engravings give a total of forty-six icons – ten more than Harvey – against Newton's total of 231 icons up to 1800.

Although forty-five years younger, Alexander Pope died only seventeen years after Newton in 1744. He sat for his portrait to be painted twelve times, as compared to thirteen sittings by Newton. With an unknown number of drawings by Jonathan Richardson, the elder (1665–1745), two busts (1730 and *c.*1760) by Michael Rysbrack (1694–1770), five busts (two in 1738, one in 1740, and two in 1741) by Roubiliac, and other sculptures, and adding the numerous copies and engravings made of the twelve oil portraits, W. K. Wimsatt (1907–1975) in his authoritative account numbers eighty-one icons of Pope, mainly done during his lifetime.[7] With fewer copies of his portraits, the total number of icons made during Newton's life is about fifty, and less than a quarter of the total of those up to 1800.

Pope sat three times to Kneller, Newton four times (**II-1**, **III-1**, **VIII-1** and **IX-1**), and both of them were portrayed twice by Charles Jervas (1675?–1739) and once by Michael Dahl (1656-1743). Jonathan Richardson (1665–1745) painted Pope three times in addition to the drawings he did of him, but made only one drawing of Newton, a copy after the third portrait by Kneller (**VIII-3**). Wimsatt wrongly considered that 'Pope was probably the most frequently portrayed English person of his generation, perhaps of the whole eighteenth century'.[7] He was writing as a literary historian and ignored the broader culture of that century. In the number of their portrayals in the eighteenth century, Pope was outpaced by Newton by nearly a factor of three. David Garrick (1716–1779), who had a marble bust of Newton carved by Peter Scheemakers *c.*1730 in his collection (**A.30**), was more often painted than anyone else in England in the eighteenth century.[8] There were sixteen portraits in oil and sixteen paintings of him in theatrical scenes done by thirteen different artists: thirteen engravings were made after the theatrical scenes, but no copies of the portraits. As well, there were ten drawings and fourteen sculptures and ceramics, bringing the total number of Garrick's icons, all done during his life-time, to seventy, which can be compared with the fifty icons of Newton made during his life and the total of 231 up to 1800.

The *ad vivum* portraits and their analysis

Joseph Addison (1672–1719) wrote on the first page of the first issue of *The Spectator* in 1711:

> I have observed that a reader seldom peruses a book with pleasure 'till he knows whether the writer of it be a black or a fair man, of a mild or choleric disposition, married or a bachelor, with other particulars of a like nature, that conduce very much to the right understanding of an author.

And Jonathan Richardson remarked in his introduction to *An Essay on the Theory of Painting*, 1715:

> Painting gives us not only the persons, but the characters of great men. The air of the head and the mien in general give strong indications of the mind and illustrate what the historian says more expressly and particularly. Let a man read a character in my Lord Clarendon (and certainly never was there a better painter in that kind), he will find it improved by seeing a picture of the same person by Van Dyck.[13]

Which remarks by Richardson may be compared with those by Johann Kaspar Lavater (1741–1801) in his *Physiognomische Fragmente*, vol. IV (Leipzig, 1778), in a discussion on four engraved portraits of Newton made for his book by Johann Lips (1758–1817), which is included in an Appendix to Part I.

Newton's appearance can be well discerned from the authentic *ad vivum* portraits illustrated in this book. The feeling that the faces in the differing presentations sometimes show a worrying variability has surely derived from the wrong acceptance of other portraits, such as those found to be spurious and doubtful (**XXXII** to **XLII**).[14] No allowance should need to be made for the varying skills of the portrait painters, nor for their aims in making the paintings, aims that include the reason for their commissioning. In only three of the fourteen portraits is this not known, or it is not possible to make a safe surmise why it was done (**V-1**, **VII-1** and **X-1**). The disguising effect in half of the portraits of the wearing of a full-length wig, and the changes with age, which show Newton becoming increasingly plump, can be ignored. After comparing the variation in his authentic portraits with the variation in portraits of others in the eighteenth century, the idea that in their commissioning Newton unduly manipulated his own appearance, as suggested by Fara, is not tenable.[15]

It is of interest to analyse the fourteen portraits in oil done from life. In seven of them Newton is wigged and thus more formal in appearance. Of these, four were commissioned, including three of those by Kneller (**III-1** for Newton, **VIII-1** for the Earl of Macclesfield and **IX-1** by Newton to send to Pierre Varignon in Paris), and the portrait by Jervas presented by Newton to the Royal Society (**VI-1**). In addition, the portrait with a wig by John Vanderbank (1694?–1739), done in the last months of Newton's life when he was already unwell, was commissioned by Martin Folkes (1690–1754) for probable eventual presentation to the Royal Society (**XV-1**). The portrait by Thomas Murray (1663–1735), which was owned by a half-niece (**VII-1**), and the unattributed portrait **X-1** also have Newton wearing a wig.

In the portrait by Kneller of 1689, commissioned by Newton when he was in London representing the University of Cambridge in Parliament (**II-1**), which he kept to the end of his life (as he did **III-1**), he is unwigged, as he is in the portrait by Sir James Thornhill (1675–1734) commissioned by Richard Bentley (1662–1742) and given to Trinity College,

Cambridge (**IV-1**). He does not wear a wig in the portrait by Christian Richter (1682?–1732), which is only known by a poor engraving of it (**V-1**), nor in the indifferent drawings by William Stukeley (**XI-1**, **XI-2** and **XI-3**). He is unwigged in three heads: 1. (**XII-1**) by Michael Dahl (1656–1743) perhaps commissioned by 2nd Earl of Oxford (1689–1741); 2. (**XIII-1**) by Vanderbank, with the portrait **XIII-2** perhaps commissioned for the frontispiece for the 3rd edition of *Principia*, 1726; and 3. (**XIV-1**) by Enoch Seeman (1694–1744), 1726, with the portrait **XIV-2** presumably commissioned by Robert More, FRS (1703–1780). In the bust-profile of Newton in the background of the portrait of Samuel Clarke (1675–1729), after Jervas, commissioned by Lord Lechmere (1675–1727), he also is not wearing a wig (**XVI-2**).

There are three autograph and artist's studio copies of Kneller 1689 (**II-1**), five of Kneller 1702 (**III-1**), three of Thornhill (**IV-1**), five of Vanderbank 1726 (**XIII-2**) and three of Seeman 1726 (**XIV-2**). Five portraits copied more than once by other artists were: 1. Kneller 1689 (**II-1**) twice; 2. Kneller 1702 (**III-1**) five times; 3. Thornhill (**IV-1**) three times; 4. Vanderbank 1726 (**XIII-2**) three times; and 5. Seeman (**XIV-2**) six times. The portraits engraved more than once were Kneller 1702 (**III-1**) 13 times; Vanderbank 1726 (**XIII-2**) nine times; and Seeman (**XIV-2**) five times. As already mentioned, only four of the engravings were made during Newton's life, two in 1712 from portraits **III-1** and **IV-2**, and two from **XIII-2** in 1726 in the last year of his life.

The posthumous sculptures

With the exception of the le Marchand ivories, all the sculptures were made after Newton's death. In 1782, Thomas Maude (1718–1798) wrote in the appendix to his poem *Viator*:

> Various are the effigies of Sir Isaac, both in frontispieces, medallions, busts, seals, and other engravings, but most of them dissimilar from his monument and from each other … The Rev. Dr. Birch has given us a pompous engraving of Sir Isaac in his folio volume[7] [**XLIII-2** engraved by Jacob Houbraken], which, from the attention paid it in the executive part, may have merit as to similitude.[14]

This somewhat sweeping statement fifty-five years after Newton died must mainly apply to the busts and reliefs, and to the poor likenesses in the engravings done after the *ad vivum* portraits, often by jobbing copyists. It might have included, perhaps, the engraving by George Bickham (1684–1769), first published in 1732 (**LV-1**), which Maude (who had never seen Newton) had referred to earlier, but more likely the unacceptable spurious and doubtful images listed in Part I.

John Conduitt, FRS (1688–1737) who was married to Newton's half-niece, Catherine Barton (1679–1740), and had succeeded him as Master of the Mint, a position that Newton had held for twenty-seven years, was assiduous in commissioning and advising on the icons to celebrate Newton and his achievements. In addition to the 'pictures or draughts [sketches]' of Newton that Conduitt provided, Rysbrack and Roubiliac in carving the portrait busts were able to use the ivory bust **A.2** and one of the copies of the death mask, and for Rysbrack the monument in Westminster Abbey (**D.1-D.6**) and for Roubiliac the statue in the Ante-Chapel in Trinity College, Cambridge (**D.7**).

Webb in her book[16] *Michael Rysbrack – Sculptor* finds that the face in Rysbrack's bust (**A.10**) bears a closer resemblance to the death mask than in Roubiliac 's statue (**D.7**). Newton's figure was more thickset and shorter at 5 feet 6 inches than he was made in the statue at 5 feet 11 inches tall, where he is made to look, according to Webb, 'younger and more elegant and romantic than in fact he did look'. She believes that both sculptors idealised their sitters, but Rysbrack less than Roubiliac. This flattery by idealisation, it can be added, does not explain Maude's remark, especially on reading a Minute of the Council of the Royal Society of 18 August 1785 on the bust **A.18**:

> N. B. This bust in Terra Cotta was made under the Eyes of Mr Conduitt and several of Sir Isaac's particular friends by Roubiliac, from many Pictures and other Busts, and esteemed more like than anything extant of Sir Isaac.

It should be noted that it is particularly in recording the thirty-four portrait busts made by nine different sculptors, all of them easily recognisable as of Newton, that new attributions have been made. The sculptors, with varying success of course, tried to get a good likeness, and in section E on ceramics this equally applied to the busts and portrait medallions coming from the Wedgwood factory workshop at Etruria.

A note about dates

The Gregorian new-style calendar came into general use in Catholic countries in 1585. Change was made so that each year now began on 1 January instead of on 25 March, with certain leap years eliminated and ten days subtracted to correct for this. But England continued to use the Julian old-style calendar, treating the new-style calendar as a piece of popish superstition. It was George Parker, 2nd Earl of Macclesfield (1697–1764), President of the Royal Society from 1752 to 1764, who in 1751 presented a Bill to Parliament to introduce the Gregorian calendar into the United Kingdom. The change to New Style was made in 1752, when eleven days had to be subtracted rather than ten (as England had observed 1700 as a leap year), bringing the end of the fiscal year from Lady Day, 25 March, to the present 5 April. It may be noted that contemporaries, not unexpectedly, became very confused by these changes in the method of dating.

In this iconography, the year of the icon, where known, is given in New Style, and taken to have begun on 1 January. Conventionally, Newton's birthday is kept to 25 December 1642 (instead of, in New Style, 4 January 1643), and his death to 20 March 1727 (instead of, in Old Style, 20 March 1726, or, more correctly in New Style, 31 March 1727). The years inscribed on the *ad vivum* portraits of Newton, all necessarily done before 1752, are in Old Style, so that, when Newton's age, or the month, is also given, it may indicate that the painting was made between 1 January and 25 March of the next year in New Style. Such a portrait is that by John Vanderbank which is dated 1725 on the canvas (**XIII-2**), but with his age put as '84' on the mezzotint of it by John Faber (1684–1756) (**XLVII-1**). Newton entered his eighty-fourth year on Christmas Day, 1725, which implies the portrait was painted between then and 25 March in 1726.

Bibliography

1. R. E. W. Maddison. 'The portraiture of the Honourable Robert Boyle, F.R.S.' *Annals of Science* 1959; 15: 141–214 with 55 plates.
2. Samuel Johnson. *The Idler* no. 45, 1759.
3. D. Beales. 'History and biography: an inaugural lecture, 1980' in T. C. W. Blanning and D. Cannadine (eds.), *History and Biography: Essays in Honour of Derek Beales.* Cambridge: Cambridge University Press, 1996, pp. 266–83.
4. D. E. Smith. 'Portraits of Sir Isaac Newton' in *Isaac Newton 1642–1727.* A memorial volume edited for the Mathematical Association by W. J. Greenstreet. London: G. Bell and Sons, 1927, pp. 170–8.
5. F. E. Brasch. 'Newton's portraits and statues'. *Scripta Mathematica* 1941; 8: 199-227.
6. Geoffrey Keynes. *The Portraiture of William Harvey: The Thomas Vicary Lecture for 1948.* 2nd edn. London: The Keynes Press, British Medical Association, 1985.
7. W. K. Wimsatt. *The Portraits of Alexander Pope.* New Haven and London: Yale University Press, 1965, p. xv.
8. D. Shawe-Taylor. 'The beautiful strokes of a great actor – Garrick and his painters' in *Every Look Speaks: Portraits of David Garrick*, with catalogue of exhibition. Bath: Holburne Museum of Art, 2003.
9. Thomas Birch. *The HEADS of Illustrious Persons of Great Britain engraved by Mr. Houbracken, & Mr. Vertue with their LIVES and CHARACTERS by Thomas Birch, A.M., F.R.S.* London: John & Paul Knapton, 1743, pp. 147–8, plate opposite p. 147.
10. Abraham Langford. *A Catalogue of the genuine and capital Collection of Pictures by the most celebrated MASTERS, Of that late Great and Learned Physician, Doctor Richard Mead, Deceased, will be sold by auction, at Mr Langford's, Great Piazza, Covent-Garden on 20th, 21st, and 22d of this instant March 1755.* London, 1755.
11. L. M. Principe. 'Robert Boyle's portrait at Chemical Heritage Foundation'. *Chemical Heritage* 2002; 20: 8-9.
12. Malcolm Baker. 'The making of portrait busts in the mid-eighteenth century: Roubiliac, Scheemakers and Trinity College, Dublin'. *Burlington Magazine* 1995; 137: 821-31.
13. Jonathan Richardson, the elder. *An Essay on the Theory of Painting by Mr Richardson.* London: John Churchill, 1715, introduction.
14. Thomas Maude. *Viator, a poem: or, A Journey from London to Scarborough by the way of York.* London: B. White, T. Becket and J. Walter; and the Booksellers in York, 1782, appendix p. ii.
15. P. Fara. *Newton: The Making of Genius.* London: Macmillan, 2002, pp. 28–31.
16. M. I. Webb. *Michael Rysbrack – Sculptor.* London: Country Life Ltd, 1954, p. 81.

PART I

PORTRAITS

AD VIVUM PORTRAITS

I. David Loggan (1635–1700)

I-1 John Collins, FRS (1625-1683), writing to Newton on 5 March 1677,[27] wrote that 'Mr Loggan' [who was a resident in Trinity College, Cambridge, for ten years as he prepared the engravings for his *Cantabrigia Illustrata* (1688)[21]] 'informs me he hath drawn your effigies in order to a sculpture [engraving] therof to be prefixed to a book of Light Colours Dioptricks which you intend to publish.' No drawing or engraving of Newton by Loggan has survived, and *Opticks*[26] only appeared in 1704.

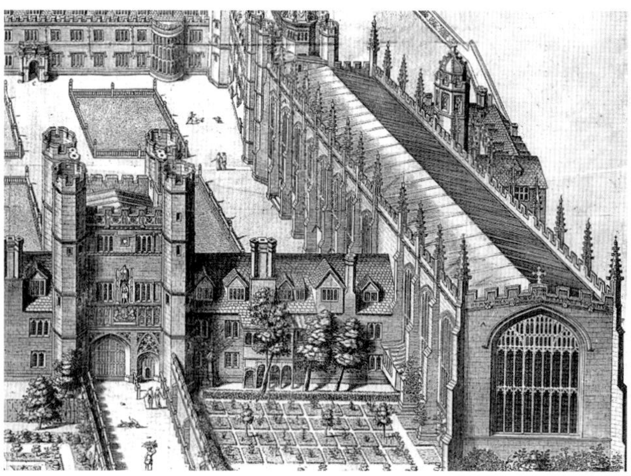

Figure 1. Part of David Loggan's engraving of Trinity College from *Cantabrigia Illustrata*, 1688. It shows Newton's first-floor rooms with balcony in the building that joins the Great Gate to the Chapel, facing Trinity Street. Below is his laboratory in the corner of his well-tended garden, lying between the building and a buttress of the Chapel.

II. Sir Godfrey Kneller (1646–1723)

II-1 Aged forty-six, oil on canvas, 1689.
73.7 × 61 cm (29 × 24 in)
Earl of Portsmouth, Farleigh Wallop, Basingstoke, Hampshire (Stewart[31] no. 528).

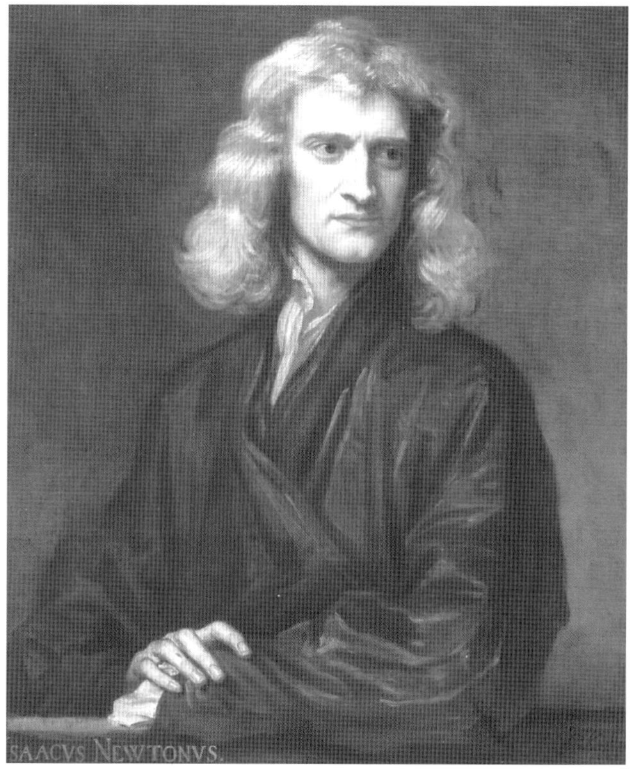

II-1
see colour Plate I

Half-length, with left arm across body, leaning on a ledge; head half-turned to right, eyes looking to his left, with long greying hair reaching to shoulders; body turned half-left, wrapped in dark gown with buttonless white shirt at neck; right hand overlaps left wrist with white cuff; dark background.

On canvas, at left base, written on the ledge:
ISAACVS NEWTONVS.
Signed and dated at right lower corner:
GKneller
1689

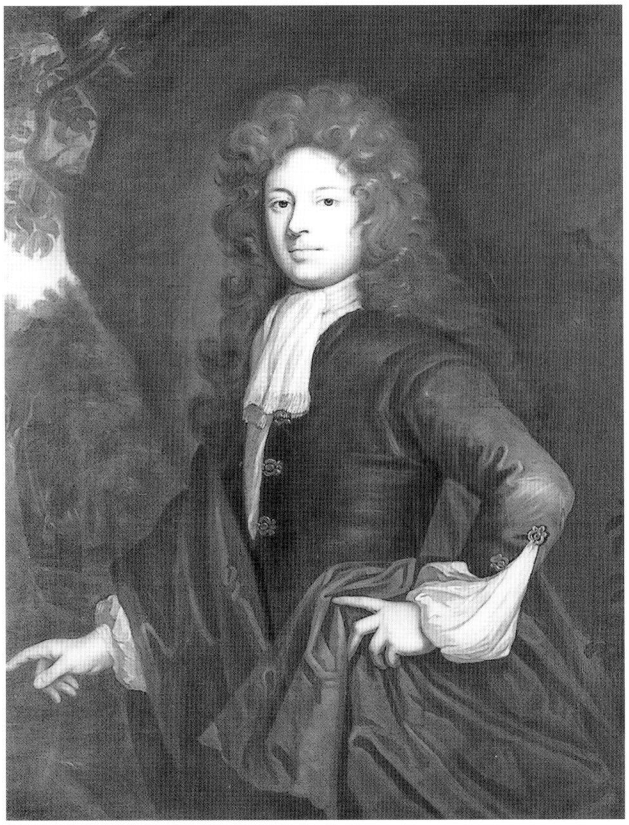

Figure 2. Portrait of Charles Montagu, later Earl of Halifax, by Sir Godfrey Kneller, 1685. (Trinity College, Cambridge)

Isaac Newton was elected as one of the two representatives of the University of Cambridge to the Convention Parliament, which first met, following the 'Glorious Revolution', in December 1688. (In Parliament he voted as a Whig and as a militant anti-papist, but only spoke once, when he requested that a draughty window be closed.)[37] His time as an MP lasted until 27 January 1690, when William III first prorogued Parliament, and then dissolved it on 6 February. He was not re-elected. He spent his time in London greatly busy with university affairs, and it was during this time that he was first painted by Kneller, as his friend, Charles Montagu, FRS (1661–1715), later 1st Earl of Halifax, had been in 1685: Trinity College purchased this portrait (above) of Montagu (oil on canvas, 123.2 × 98.5 cm ($48\frac{1}{2} \times 38\frac{3}{4}$ in)) for 17 guineas in 1852 (Bursar's Account, 1852).

The portrait **II-1** is very freely and boldly executed, and has been described by Oliver Millar[23] as showing Newton 'visibly thinking'. The inscription on **II-3**, a studio variant of **II-1**, has 'Aetatis 47' added to the year '1689', with Newton entering his 47th year on 25 December 1688. Newton kept the portrait in his possession until his death, when it passed to John Conduitt, FRS (1688–1737) and his wife, Newton's half-niece, Catherine Barton Conduitt (1679–1740). In 1740 their daughter Catherine (Kitty) (1718–1750) married John Wallop (1718–1749), afterwards Viscount Lymington when his father, also John Wallop (1690–1762), was created Earl of Portsmouth in 1743. The portrait passed to their grandson, John, 2nd Earl of Portsmouth (1742–1797), and then by descent.

II-2 Studio copy of **II-1**, oil on canvas.
75 × 62.2 cm ($29\frac{1}{2} \times 24\frac{1}{2}$ in)
The Library at The Vyne, Basingstoke, Hampshire (National Trust) by 1842.

Perhaps owned by John Chute (1701–1776), friend of Horace Walpole (1717–1797) and whose brother Francis (1696–1745) apparently knew Newton.

II-3 Studio variant after **II-1**, oil on canvas, 1689.
125.7 × 101.6 cm ($49\frac{1}{2} \times 40$ in)
Royal Collection, Hampton Court Palace. (Millar[23] no. 363)

Three-quarter-length, standing beside a table, on which is a terrestial globe with the words 'America' and 'Mare Pacificum' appearing on it, and an open book, with the name *DANIEL*, written in Greek capitals, heading its left page; head half turned to right, and eyes looking to his left, with long hair; arms lying across body, which is turned half-left, and hands separated in front; wearing a waistcoat rather than a gown, with open shirt at neck.

Inscribed on canvas:

Isack Newton Esqr. AEtatis 47 / 1689

Newton, who was not knighted until 16 April 1705, was in his forty-seventh year in 1689. This painting is

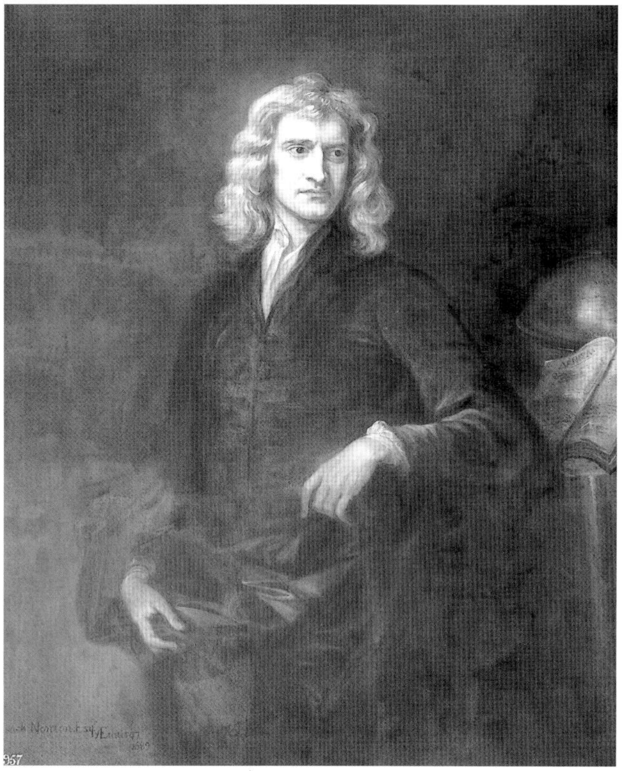
II-3

probably one of the 'Portraits of several Poets, Painters, and Philosophers', some of which may have been collected by Queen Caroline of Ansbach, recorded in the Queen's Dressing Room at Windsor, but later moved between St James's, Hampton Court and Kensington Palaces.[23] Besides **II-3**, Caroline had earlier set up portraits of Dr Samuel Clarke (1675–1729), John Locke, FRS (1632–1704), Robert Boyle, FRS (1627–1691) and William Wollaston (1660–1724) – see **XVI-2** – in Kensington Palace. Caroline (1683–1737), who first came to England as Princess of Wales in 1714, became Queen to George II in 1727, but when she may have acquired the painting **II-3** is unknown.

Newton's *Observations upon the Prophecies of Daniel and the Apocalypse of St John* was written into old age over many years beginning in the mid-1670s; the Revelation of St John the Divine was discussed with the theologian Henry More, FRS (1614–1687) in 1680. When Newton made his precise research on Daniel is not known: he made a schematic interpretation of the Book of Revelation which he sent to John Locke, whom he probably had first met in 1689, in 1690. *Obervations* was first published by his half-nephew, Benjamin Smith (1700–1776) in two parts in 1733. Whether there is any significance in the west coast of America being shown on the globe is not clear.

The portrait is poorly painted particularly in its lower third, with the right hand apparently changed from holding something rather than showing pentimenti. From a photograph Stewart finds the painting at best a studio work. The head is obviously based on **II-1**, but treated in a much harder, brighter fashion. The highlights in the hair have a mechanical quality to them. It is possible that **II-3** is a copy of an autograph Kneller, but equally possible that an early eighteenth-century artist borrowed the 1689 Kneller head. Stewart adds that the design of the pose seems flaccid, and is reminiscent of the pose of the portrait of Michael Rysbrack attributed to John Vanderbank, *c.*1728, in the National Portrait Gallery (no. 1802). He does not know of a Kneller pose quite like it.

II-4 Unknown artist after **II-1**, oil on canvas.
89.5 × 68.5 cm ($35\frac{1}{4} \times 27$ in)
Collection: Viscount Galway, Serlby Hall, Notts.; Christie's 23 March 1979 (lot 124); Martayan Lan, New York, 2004.

II-5 Unknown artist, drawing after **II-1**, in pen, brown ink and grey wash.
Oval, 14 × 10.8 cm ($5\frac{1}{2} \times 4\frac{1}{4}$ in)
Pepys Library, Magdalene College, Cambridge (2979/126c).

The drawing is stuck down with other small portraits in a large album and below it is written: 'Mr Newton'. Samuel Pepys (1633–1703) was President of the Royal Society from 1684 to 1688, and it was his name that was placed on the title page of *Principia*[25] underneath the Society's imprimatur of 5 July 1686. He was in particular contact with Newton at the time of the latter's breakdown in 1693, following on with a correspondence on chance in the throwing of dice.[27]

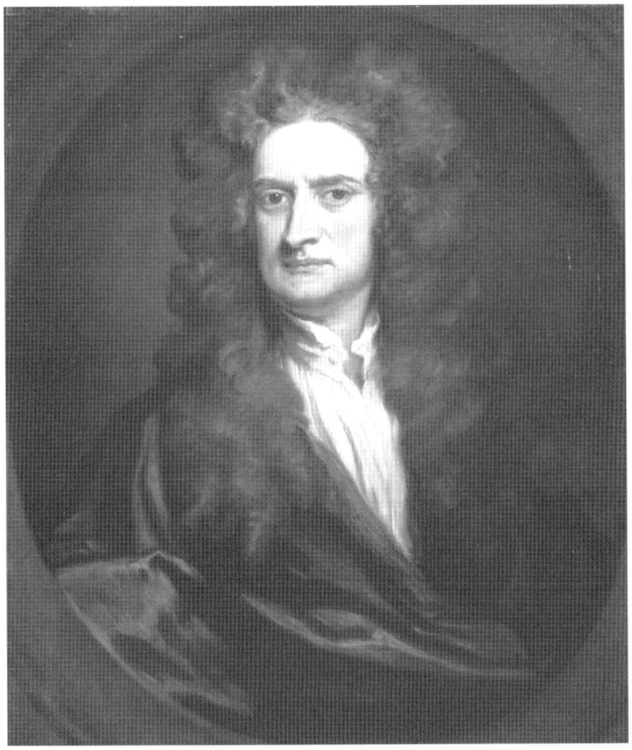

III-1
see colour Plate II

III. Sir Godfrey Kneller (1646–1723)

III-1 Aged fifty-nine, oil on canvas, 1702.
75.6 × 62.2 cm ($29\frac{3}{4} \times 24\frac{1}{2}$ in)
National Portrait Gallery (Piper[29] no. 2881; Stewart[31] no. 529).

Head and shoulders, half-length, in feigned oval; head turned slightly to left, shoulders to right; large blue-grey eyes looking front; brown wig falling below the shoulders; white shirt with collar unbuttoned, no cravat; reddish brown drapery about the shoulders.

Signed and dated left centre:

GKneller 1702 (GK in monogram)

Newton kept this portrait in his possession until he died, when by descent from John Conduitt to the Earls of Portsmouth; Viscount Lymington Sale of The Newton Papers at Sotheby's, 14 July 1936 (lot 330), when bought by the National Portrait Gallery with the aid of the National Art Collections Fund. An inscription (lower left) 'Sr ISAAC NEWTON/ Painted by Sir Godfrey Kneller/ 1702' on the canvas was removed in 1983 when the portrait was cleaned. The head of the posthumous portrait (**XIX-1**) by **Thomas Hudson** at Trinity College, Cambridge, is based on this portrait, which was engraved thirteen times before 1800, the first, a mezzotint (**XLIII-1**) by **John Smith**, in 1712.

III-2 Studio or autograph replica of **III-1**, oil on canvas.
76.2 × 63.5 cm (30 × 25 in)
Hunterian Museum and Art Gallery, Glasgow University.

On canvas, in the lower left background of the oval, at the sitter's right shoulder:

GKneller / f. (GK in monogram)

Provenance: Dr Richard Mead, FRS (1673–1754), Newton's physician; bought by the dealer Oram at the third day's sale of Mead's collection of pictures at Mr Langford's, Great Piazza, Covent-Garden, on March 22, 1755 (lot 34);[20] William Hunter, FRS (1718–1783), who bequeathed the portrait to Glasgow University.

III-3 Studio copy after **III-1**, oil on canvas.
73.7 × 61 cm (29 × 24 in)
Royal Mint, London Gallery (Government Art Collection GAC 6732).

Bust-length, facing slightly to left, with eyes looking forward, in full wig reaching to mid-chest; body half-right draped in red-brown coat, and tied white cravat with unbuttoned shirt.

On canvas, upper right:

Sr. I Newton

This portrait passed down in Newton's family through Catherine Barton Conduitt's sister, Margaret Barton Warner (b.1687). Her great-granddaughter, Harriet Warner, married John Sherman Elliott, and, after their grandson C. H. B. Elliott of Frenchay, Bristol, died in 1939, his son sold the painting; Earl of Lucan; Sotheby's 21 October 1964 (lot 3) when bought by the Ministry of Works for the Royal Mint.

III-4 Studio copy after **III-1**, oil on canvas.
76.2 × 63.5 cm (30 × 25 in)
Lord Sackville, Knole Park, Sevenoaks, Kent.

High on right of canvas:
SR. ISAAC NEWTON. Obt. 1727.

Inscribed on back:
Sir Isaac Newton, painted by Sir Godfrey Kneller.
Purchased by His Grace John Fredk,
[3rd] Duke of Dorset [1745–1799], in 1777.

III-5 Studio copy after **III-1**, oil on canvas.
76.2 × 63.5 cm (30 × 25 in)
Historical Portraits, London.

Provenance: The Vicarage, Burley-on-the-Hill, Rutland [12 miles from Woolsthorpe] (label on back of frame); Phillips Bayswater sale, 8 March, 2001.

III-6 David Paton (*c.*1650–*c.*1710)

Miniature, after **III-1**, grey wash on vellum.
Oval, 9.7 cm ($3\frac{13}{16}$ in)
Sotheby's 7 March 1983 (lot 22).

Inscribed on backing paper: 'This original Drawing of/ Sir Isaac Newton belonged formerly to Professor Gregory/ of Oxford [David Gregory, FRS (1661-1708)]; By him it was/ bequeathed to his youngest/ son Isaac (Sir Isaac's Godson)/ who was later Secretary/ of Sion College; by him left/ by will to Revd. Mr. Mence,/ who had the goodness to give/ it to Dr. Douglas [? Dr James Douglas, FRS (1675–1742)]; March 8th/ 1780'

Paton was a miniaturist mainly working in Scotland and Italy, and whose self-portrait (no. A677), 16.7 × 13.9 cm, pencil on paper, signed and dated: 'David Paton Scozzese Pict. Flor. 1683' is in the Uffizi Gallery.[12,38] In an archive in Florence, there is a letter dated 8 June 1708 written to Cosimo de' Medici III, Grand Duke of Tuscany (1642–1723) from his agent in London saying that he was sending to him that day a portrait of Sir Isaac Newton, a gift from the elderly David Paton, and that he hoped that Cosimo finds no diminution in this work from the strengths and boldness of Paton's earlier years.[12] There is no other record of this second miniature of Newton by Paton.

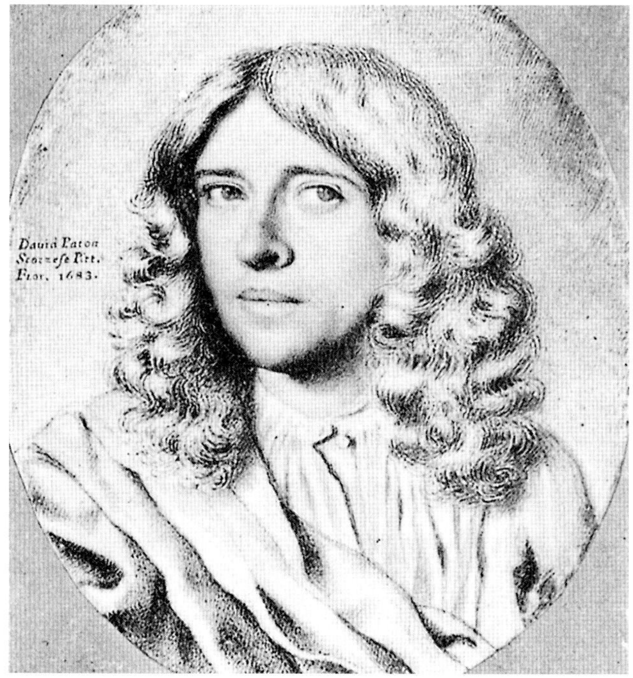

Figure 3. Self-portrait, miniature pencil drawing, by David Paton, 1683. (Uffizi Gallery, Florence)

III-7 Portrait in reverse by unknown artist from a mezzotint by John Smith of 1712 (**XLIII-1**), after **III-1**, oil on canvas.
60 × 46 cm ($23\frac{5}{8} \times 18\frac{1}{8}$ in)
Vasari Corridor, Uffizi Gallery, Florence (no. Ic 341).

Bust-length, full face with head turned slightly to right, but eyes forward; wearing wig reaching to shoulders; white shirt unbuttoned at neck with dark coat buttoned at front. Across top is written:
ISAACVS NEWTO^{N}

There is also in the Uffizi a portrait (no. Ic 77) of Robert Boyle, FRS (1627–1691), 60 × 46 cm, oil on canvas, the same size as Newton's with the same lettering to his name across its top, by an unknown artist, after an engraving by François Diodati (1647–1690).[38] Both form part of the Collezione Gioviana, consisting of more

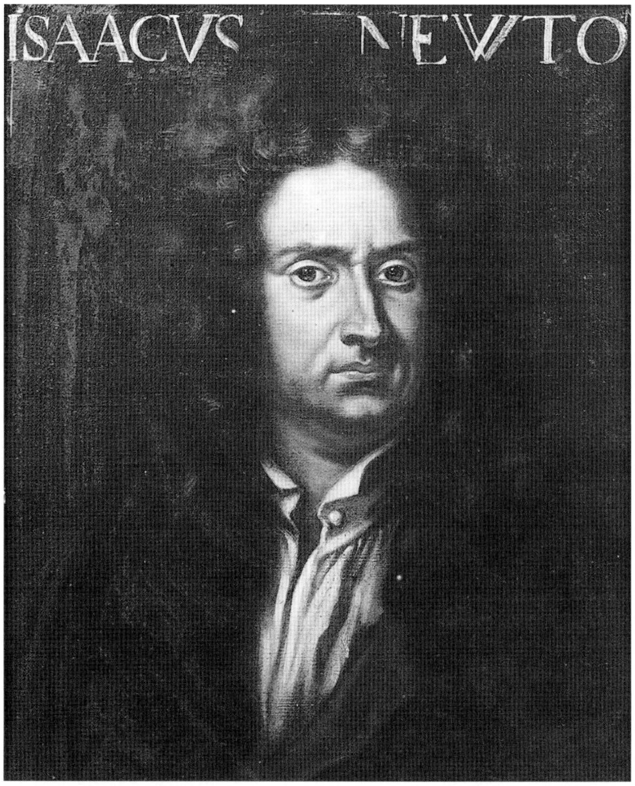

III-7

than five hundred portraits of illustrious personages, which was begun by Cosimo I (1519–1574), and added to by successive Grand Dukes. Edward Wright, who was the tutor to George Parker (1697–1764), later 2nd Earl of Macclesfield, on the Grand Tour during the years 1720–2, mentions both portraits in his description of the Palazzo Vecchio in 1722: 'On the Walls, over the Statues, are ... Persons of all Nations that have excelled in Arms or Arts. There is the *Duke of Marlborough* [1650–1722] (two of Him), Sir *Isaac Newton*, Dr *Wallis* [John Wallis, FRS (1616–1703), 61 × 47 cm, no. Ic 480], Mr *Boyle*, Mr *Ray* [John Ray, FRS (1627–1705), 61 × 47 cm, no. Ic 384] and several others of our Nation.'[38]

Prince Cosimo (later Cosimo III) visited England for ten weeks in 1669, and was given a warm reception by Charles II and his court and by the universities of Oxford and Cambridge, chiefly, it was reasoned, because his father, Ferdinand II (1610–1670), was supposed to have protected Galileo Galilei (1564–1642) against papal persecution.[38] When the Prince visited Cambridge on 1 May 1669, he met 'The More Distinguished People at the University', listed as James Fleetwood (1603–1683), Vice-Chancellor and Provost of King's College, Peter Gunning (1614–1684), Master of St John's College, John Pearson (1613–1686), Master of Trinity College, Ralph Cudworth (1617–1688), Master of Christ's College, William Sancroft (1617–1693), former Master of Emmanuel College, Archdeacon and later Archbishop of Canterbury, Edward Rainbowe (1608–1684), former Master of Magdalene College, Bishop of Carlisle, Henry More, FRS (1614–1687) and Henry Jenkes, FRS (d.1697).[38] Newton, not in the list, possibly was among those who lined the Prince's walk to the Schools, while wearing his new MA gown[37] – he had become Master of Arts and a Major Fellow of Trinity College the year before, but only later was appointed Lucasian Professor, on 29 October 1669. And he may have met the Prince that evening in Trinity when the scholars performed a Latin comedy in front of him.[38] Pepys wrote in his *Diary* of 'The Prince of Tuscany, who comes into England only to spend money and see our country ... seems a very jolly and good comely man.'[38]

The Prince met Boyle, Wallis and Ray in London,[38] but the inclusion of their portraits in the Grand Ducal collection is more likely to have followed from the activities of the Accademia del Cimento, founded in 1657 by Ferdinand II's brother, Prince Leopold, but dissolved in 1667 when Leopold became a cardinal.[37] Prince Cosimo's trip to England in 1669 is unlikely to have influenced the acquisition of Newton's portrait, as Newton only became Fellow of the Royal Society three years later on 11 January 1672 after he had shown the Society his reflecting telescope, with his first paper on the nature of light and theory of colours published in *Philosophical Transactions* on 19 February. Edmond Halley, FRS (1656–1742) sent a presentation copy of *Principia Mathematica*,[25] published in July 1687, to the Florentine Vincenzo Viviani (1622–1703), a geometrician, pupil and biographer of Galileo, member of the Accademia del Cimento, and elected FRS in 1696. And we know Viviani showed his copy of *Principia* to Cosimo III from a letter in the Royal Society written to Halley from Florence on 2 February 1688.[38] Cosimo was sent David Paton's miniature of Newton (see **III-6**) in June 1708, but **III-7** must date after 1712.

III-8 Bernard Lens, the younger (1682–1740)

Miniature, after **III-1**, oil on canvas, 1708.
Oval, 8.2 × 6.2 cm ($3\frac{3}{16} \times 2\frac{7}{16}$ in)
Collection: Marquess of Bristol at Ickworth House, Bury St Edmunds, Suffolk; Sotheby's 26 November 1992 (lot 370).

Head and shoulders, with a brown wig, wearing a blue-lined day gown and open white chemise.

Signed with initials and dated: 'BL '08'.
Inscribed verso:

Sir Isaak Newton
B. Lens Fecit 1709.

III-9 Thomas Forster (*fl.*1695–1712)

Miniature, after **III-1**, graphite drawing.
Oval, 9.5 cm ($3\frac{3}{4}$ in) high
Sold Sotheby's 27 November 1972 (lot 94).

Head and shoulders, full face, wearing full wig.

III-10 Unknown artist

Head and shoulders in an oval, after **III-1**, pastel, 1716.
18.7 × 12.7 cm ($7\frac{3}{8} \times 5$ in)
Newhailes, Musselburgh (National Trust for Scotland).

Bears contemporary inscription:

Isaac Newton 1716.

Probably possessed by Sir David Dalrymple (d.1721), 1st Baronet of Hailes, who bought Newhailes and transformed it between 1717 and his death. It is in an old frame in a corridor with other portrait engravings and landscape ink drawings of 1739 and 1742 by Sir David Dalrymple (1726–1792), 3rd Baronet, later Lord Hailes.

IV. Sir James Thornhill (1675–1734)

IV-1 In his sixty-ninth year, oil on canvas, 1709–11.
127 × 99 cm (50 × 40 in)
Master's Lodge, Trinity College, Cambridge.

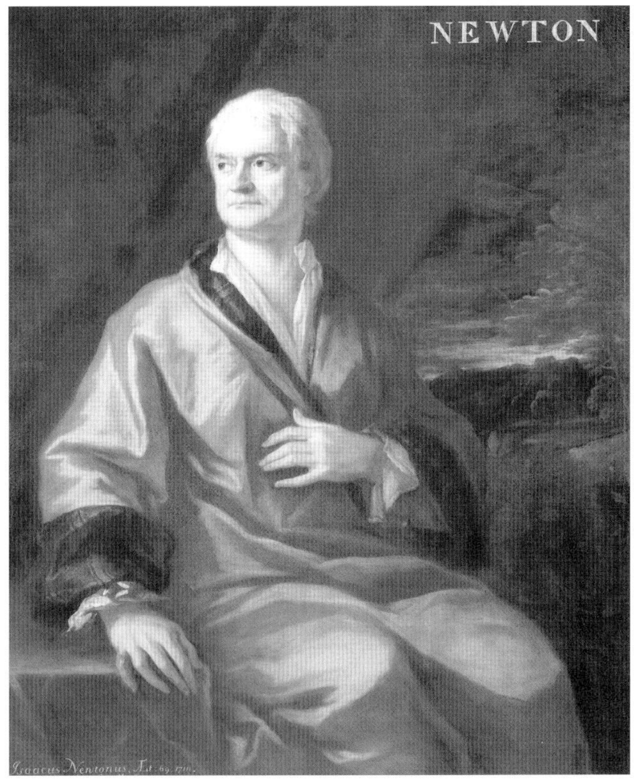

IV-1
see colour Plate III

Three-quarter-length, life-size seated figure. Head half to left, with eyes to left, and short grey-white hair; body half-turned to the right; right hand rests on red-covered table, left forearm and hand lying across body; wears light brown robe with dark collar and cuffs over a white shirt showing at wrists and open at neck. Right background has a dark landscape with a tree, mountains and clouds.

On canvas high right:

NEWTON

On canvas low left in earlier writing:

Isaacus Newtonus Aet: 69. 1710.
Donavit Bentleius Coll. Mag.

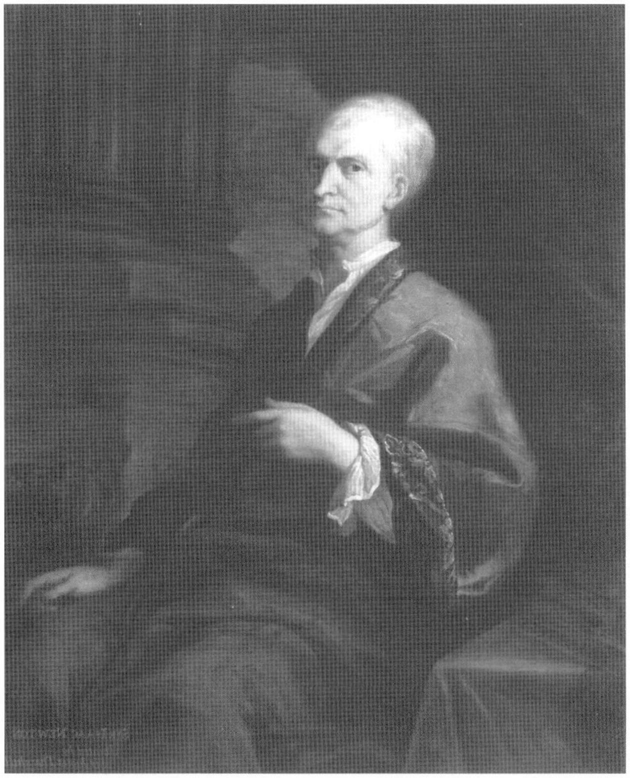

IV-2
see colour Plate III

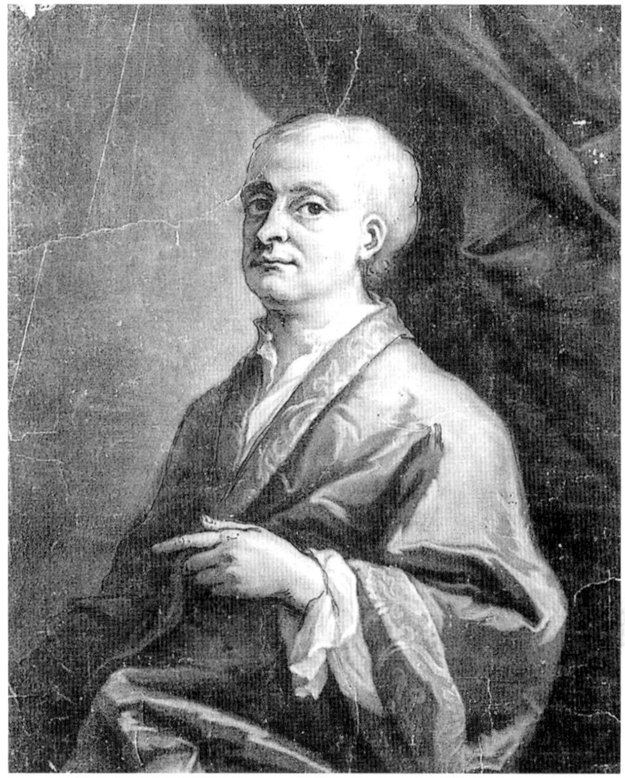

XLV-1

Commissioned and given to Trinity College by Richard Bentley, FRS (1662–1742), Master 1700–42.[11] In a letter to Newton on 20 October 1709,[25] Bentley wrote: 'I hope my Picture at Thornhills will have your last Sitting, before you leave the Town: the time you set under your Hand is already lapsed'. The inscription on the canvas implies the portrait was finished before 25 March 1711, but after Christmas Day 1710 when Newton entered his sixty-ninth year. Esdaile[14] thought that the portrait seemed to have influenced Roubiliac with his marble bust (**A.24**) in the Wren Library and statue (**D.7**) in the Chapel at Trinity College. At the time she wrote, in 1928, the terracotta bust at the Royal Greenwich Observatory (**A.18**) was not known.

IV-2 Compositional variation of **IV-1**, oil on canvas.
124.5 × 95.2 cm (49 × 38½ in)
Earl of Portsmouth, Farleigh Wallop, Basingstoke, Hampshire.

Three-quarter-length, life-sized seated figure. Head and body turned three-quarter to right, with eyes looking towards spectator, and short, grey-white hair. To the sitter's right there is a table with two books resting on it, with OPTICKS on top of PRINCIPIA. Right forearm lying horizontally across body with forefinger pointing to left; left arm rests on left thigh. Wears yellow-brown robe, with cuffs and collar of green silk with a gold pattern on it, over a white shirt open at neck. Right background has a double pillar on a base.

On the canvas, at base of pillars, is written in black:

Is: NEW
I. Tho

Low, on right, is written:

SIR ISAAC NEWTON
Painted by
Sir James Thornhill

By descent from John Conduitt to the Earls of Portsmouth. The face is thinner and more severe than in **IV-1** with its more powerful-looking jaw. There is no engraving of **IV-1**, but **Jean Simon** engraved a mezzotint

(**XLV-1**), half-length figure, after **IV-2**, which is mentioned in a letter from Roger Cotes, FRS (1682–1716) to Newton on 20 July 1712:[27] 'I thank you for the Picture which I received of him [Dr Bentley]: 'tis much better done than the former [referring to the mezzotint engraving (**XLIII-1**) by **John Smith** after **III-1**, the portrait by Kneller in 1702]; but I could have wish'd it had been taken from the first of Mr Thornhill's [**IV-1**].' It was common practice then to give an engraved portrait of yourself to your friends. Thornhill was knighted in 1720, so the inscription on the canvas must date after that: an earlier inscription has been partly removed.

IV-3 Autograph copy of **IV-2**, oil on canvas.
122 × 101.6 cm (48 × 40 in)
Woolsthorpe Manor, Grantham, Lincolnshire (National Trust) since 1995.

This portrait is differed from **IV-2** by having Newton sitting on a red cushion with gold tassles, but without the table with two books on it, nor is there any writing on the canvas.

Provenance: By descent to Sarah Newton who married John Foulds of Nottingham in 1744. Sarah was the daughter of John Newton (b.1672) of Nottingham by Sarah Bridges, and granddaughter of Robert Newton (1629–1702), a barrister of Micheleover, Derby, but her exact relationship to Isaac Newton has not been established. By descent in the Foulds family to Dorothy Hylda Foulds (1901–1977); private collection; Historical Portraits Ltd, London.

IV-4 Unknown artist. Bust-length portrait, after **IV-2**, oil on canvas.
69.3 × 56.2 cm ($27\frac{1}{4} \times 22\frac{1}{8}$ in)
Wellcome Library, London.

IV-5 Valentine Ritts (*fl.*1710–60)

Portrait after **IV-1**, oil on canvas, 1735.
225.5 × 137.2 cm (89 × 54 in)
Trinity College, Cambridge.

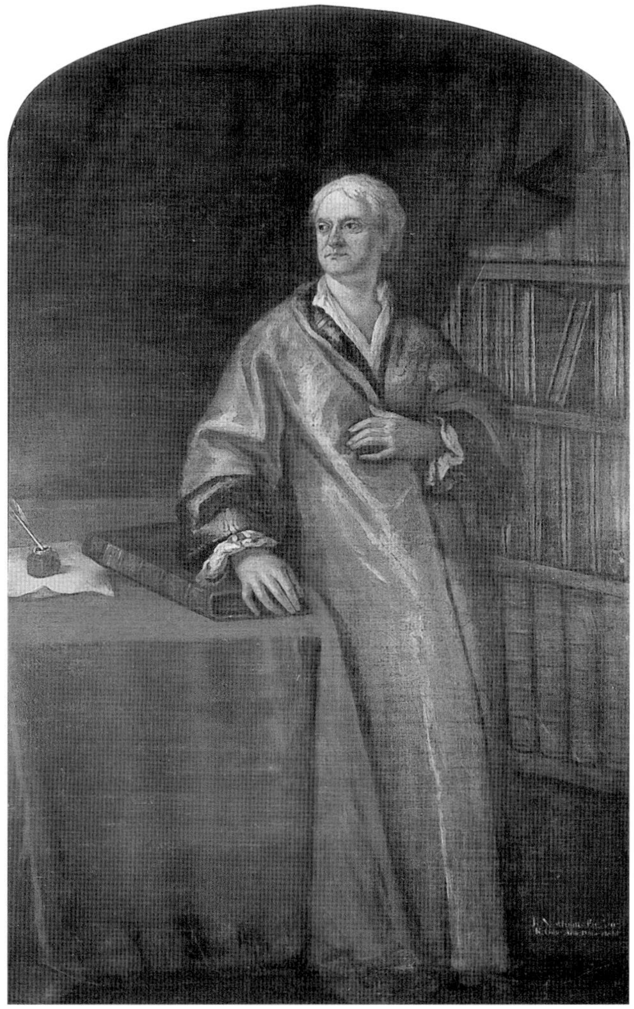

IV-5

Whole-length, life-size standing figure. Head three-quarter left, short white hair; body slightly right, standing at a table covered with a red tablecloth on left; right hand on book, with NEWTONUS written on its label, lying on table; left hand on chest; wears yellowish-brown robe, with black lining and black cuffs, over white shirt open at neck; at left, besides the book on the table, there is paper and an ink bottle with a quill standing in it; on right, there is a bookcase with five filled shelves.

At foot of canvas is written:

Is Newtonus Eq. Aur.
R. Gale Arm Dono Dedit.

below which, but with a gap, is:

Val Ritts pinx 1735

At top of frame:

Isaacus Newtonus. Eques Auratus

At foot of frame:

Ex Dono S. Knight 1752

Provenance: Roger Gale, FRS (1672–1744), Fellow of Trinity College, 1697; Samuel Knight (1718–1790), Fellow of Trinity College, 1741.

'Mr Retts', the painter, was apparently a German resident in Cambridge, who lived near Little St Mary's Church for fifty years or more, and was mainly a copyist. The painting is an indifferent copy.

As described by G. M. Trevelyan, FRS (1876–1962), Master 1941–51, in his historical sketch of Trinity College (Cambridge University Press, 1943), from when it was painted **IV-5** hung at the centre of the north end of the dining hall between portraits of Sir Francis Bacon (1561–1626) and Isaac Barrow (1630–1677). This was until 1912 when the portrait of Henry VIII, after **Hans Holbein (1497–1543)** and painted by **Hans Eworth (*fl.*1540–1573)**, was transferred from the Master's Lodge where it had hung since being commissioned sometime between 1561 and 1567. Twelve other large portraits, mainly painted in the 1730s to 1750s, also hung in the Hall, including those of Richard Bentley, FRS (1662–1742), Sir Edward Coke (1552-1634), Abraham Cowley, FRS (1618-1667), John Dryden, FRS (1631–1700), Sir Thomas Parker, FRS, 1st Earl of Macclesfield (1666?–1732), Lord Chancellor (1718–25) and pall-bearer at Newton's funeral in Westminster Abbey, and John Wilkins, FRS (1614–1672), Master from1659 till deprived of the mastership in 1660 at the Restoration, first secretary of the Royal Society, 1662–8.

IV-6 John Downman (1750–1824)

Drawing after **IV-1** in black, red and white chalk, with stump on brown paper, some lines incised with a stylus, 1777.
Oval, 22.8 × 19.1 cm (9 × $7\frac{1}{2}$ in)
Fitzwilliam Museum, Cambridge (no. 1952).

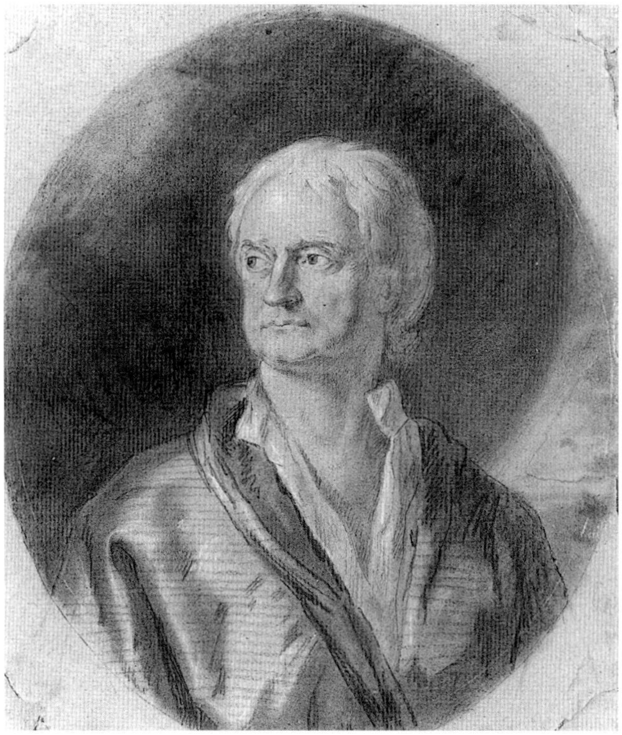

IV-6

Bust turned slightly right, facing and looking half left; bare-headed, own hair; clean shaven; wears a loose robe over a shirt open at the neck.

Inscribed below:

Sir Isaac Newton / done from the whole-length by
Sir James Thornhill in / Trinity College Hall
Cambridge 1777

This drawing occurs among a large number of original studies from life which Downman retained in his possession and used in the execution of portraits in oils. They were mounted by him in volumes arranged in five series: this drawing comes from Series II, vol. V, 30.[24] The inscription is incorrect in describing Thornhill's portrait (**IV-1**) as whole-length.

Provenance: Downman's daughter, Isabella; bought sometime before 1825 by Hon. George Neville and then passed by descent to Lady (Marjorie) Longmore, from whom acquired for the Fitzwilliam Museum in 1936 with the aid of the National Art Collections Fund.

V. Christian Richter (1682?–1732)

V-1 Aged seventy-one, oil on canvas, 1714.
12.5 × 8.8 cm ($4\frac{7}{8} \times 3\frac{1}{2}$ in)
Bust half-length, facing to right, with head turned to his right side with shoulder-length hair, wearing cravat and gown.

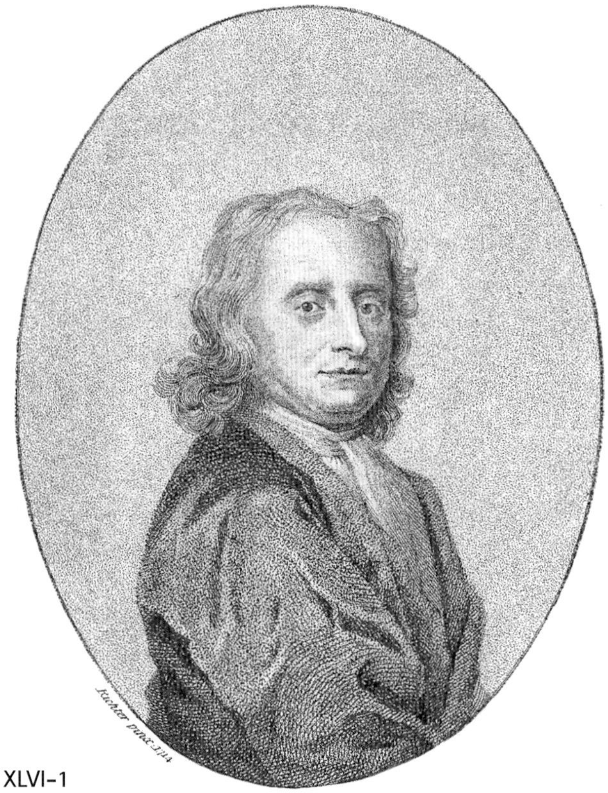

XLVI-1

The original painting is lost, but survives in an engraving (**XLVI-1**) inscribed: 'Richter pinx. 1714. Sir Isaac Newton. from a Miniature picture of the same size. London … 1785'.[40] The engraving gives a singularly poor likeness of Newton.

VI. Charles Jervas (1675?–1739)

VI-1 Aged seventy-four, oil on canvas, 1717.
125.7 × 101.6 cm ($49\frac{1}{2} \times 40$ in)
Royal Society, London.

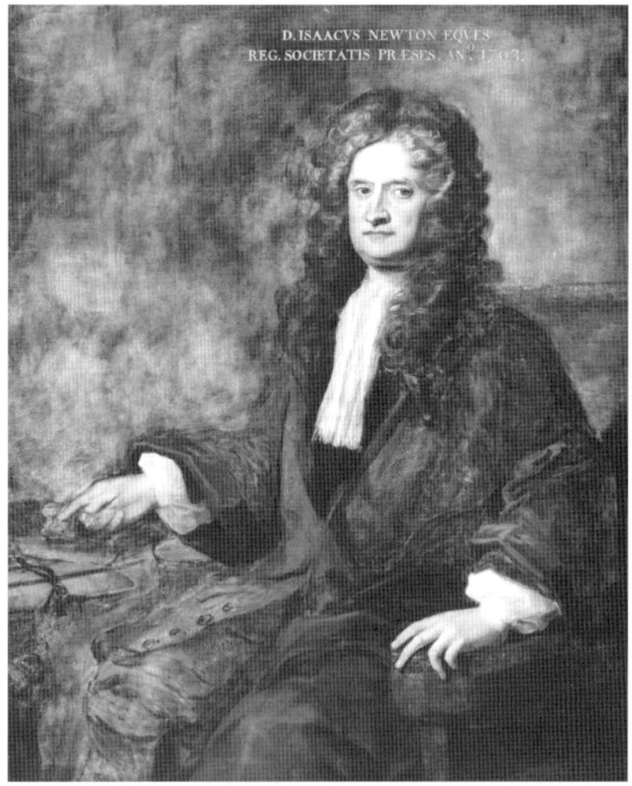

VI-1
see colour Plate IV

Three-quarter-length, seated figure, with body facing half-left; head slightly turned to left, with eyes looking to the right, wearing a full-length wig; left hand on arm of high-backed chair, and right hand with forefinger pointing at a gold watch on the table to the right side of chair; a black tape is looped through the bow of the watch and dangles over the edge of the table, with a small seal and a key, not that of a watch, attached to it; lying flat on the table below the pointing index finger is a book, with an inkwell on its cover. Newton wears a red-brown coat, buttoned at the waist, over a darker garment, a soft white cravat tied under the chin with its equal ends reaching half-way to the waist, and white cuffs; the coat covers the left leg, but the right leg in breeches is exposed.

On canvas, above head:

D. ISAACUS NEWTON EQUES
REG. SOCIETATIS PRAESES. ANo. 1703.

and on left, on side of table, is written:

Isaac Newton
1717
C.J.

The left hand is better painted than the right hand, and there is a long dark brush stroke along the top of the pointing right forefinger, suggesting something has been painted out. It would seem that Jervas originally may have painted the right hand holding a quill-pen, held above a sheet of paper on the table, later changed with a few strokes into the book, and repainting could explain the curious placing of the inkwell on the cover of the book. The watch, as may be suggested by the sitter's pointing index finger, was perhaps included as a conventional *memento mori* – a reminder of the shortness of life. The date of the repainting is uncertain, but the inclusion of the watch with the modification of the hand are surely likely to have been made at the sitter's direction.[37] An X-ray in June 2000 did not confirm that any changes by repainting had been made.

In the portrait Newton is wearing an old-fashioned wig and is painted with a face younger than his age of seventy-four in 1717, the year the painting was inscribed and dated by the artist. The portrait, commissioned by the sitter, was presented to the Royal Society at its meeting on 16 May 1717, as recorded in the Journal Book under that date: 'The President gave the Society his picture drawn by Mr Jervas for which he had their thanks.' The upper inscription on the canvas is to be read as 'President of the Royal Society [since] 1703', and not as the date of the painting as assumed by R. S. Westfall in his biography, *Never at Rest*, in 1980.[37] The gold watch in the portrait is not the one, wrongly called 'Newton's', that was bequeathed to the Royal Society by Edmund Turnor, FRS (1755?–1829) in 1829, as this has a hall-mark of 1749, a bogus inscription on its inner case, and a dial of 1810.[37]

VII. Thomas Murray (1663–1735)

VII-1 Aged seventy-five, oil on canvas, 1718.
124.5 × 100.3 cm (49 × 39½ in)
Trinity College, Cambridge.

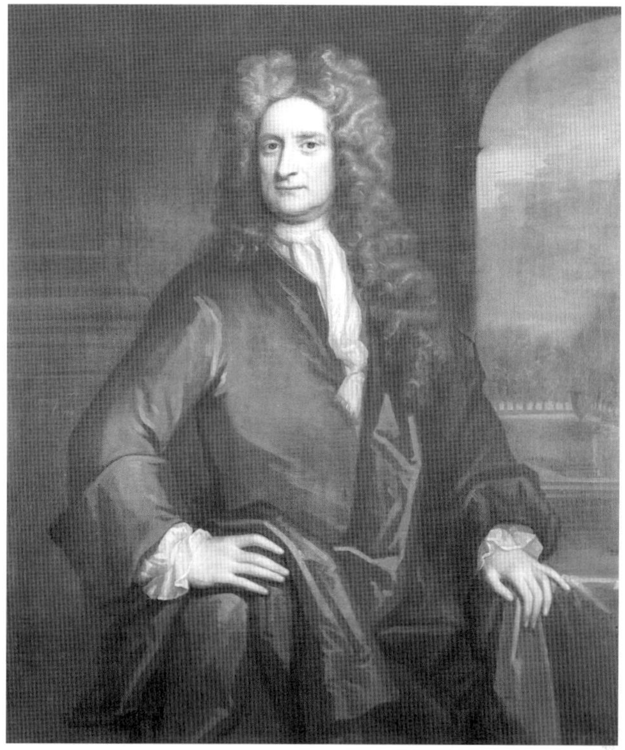

VII-1
see colour Plate II

Three-quarter-length, standing figure nearly life-size; head slightly left, eyes to front, long and full light brown wig; body half-turned to right; right hand on hip, left hand on table; wears loose dark brown coat, soft white cravat and cuffs. The archway in the right background shows a landscape.

On canvas, left centre:

T Murray F^{t}./1718

A copy of a letter, dated 23 December 1813, from Revd Charles Simeon (1759–1836), Fellow of King's College, to William Lort Mansel, DD (1753–1820), Master of Trinity College 1798-1820 and Bishop of Bristol from 1808, quotes the words of Mrs Ring of Reading, who gave the picture to Trinity College: 'My grandmother was Sir Isaac's neice [? Margaret Barton Warner (b.1687)]. The picture was taken for her and handed immediately to me. My father died when I was only fourteen days old. My mother married again in about a year and I being the only child was removed and placed under guardians. The Personals (this portrait among them) were sent to the Bank of England and did not see the light till I married at nineteen.' Simeon added that Sir Isaac had been dead at least forty years when Mrs Ring's guardians took this precaution relative to his picture. See *Charters, Wills etc, Trinity Colledge*, p. 342.[11]

A portrait of Newton by Thomas Murray (Lot 89, 7 May, p. 5) is in the catalogue at the Wellcome Library of the sale 'of the genuine and curious collection of mathematical instruments, gems, pictures, bronzes, busts, urns ... of Martin Folkes, Esq. [Martin Folkes, FRS (1690–1751)] . . . sold by auction, by Mr. Langford ... on 7th and 8th May 1755' – see **XV-1**. It is possible that this painting is the one that came into Mrs Ring's possession. Murray painted a portrait of Edmond Halley, FRS (*c.*1713), Lot 86, 7 May, p. 5 in the Folkes sale, now in the Bodleian Library, and one of Sir Hans Sloane, FRS (1725–8), now in the Royal College of Physicians.

VIII. Sir Godfrey Kneller (1646–1723)

VIII-1 Aged seventy-six, oil on canvas, 1719.
124.5 × 101.6 cm (49 × 40 in)
Private collection by descent (not in Stewart[31]).

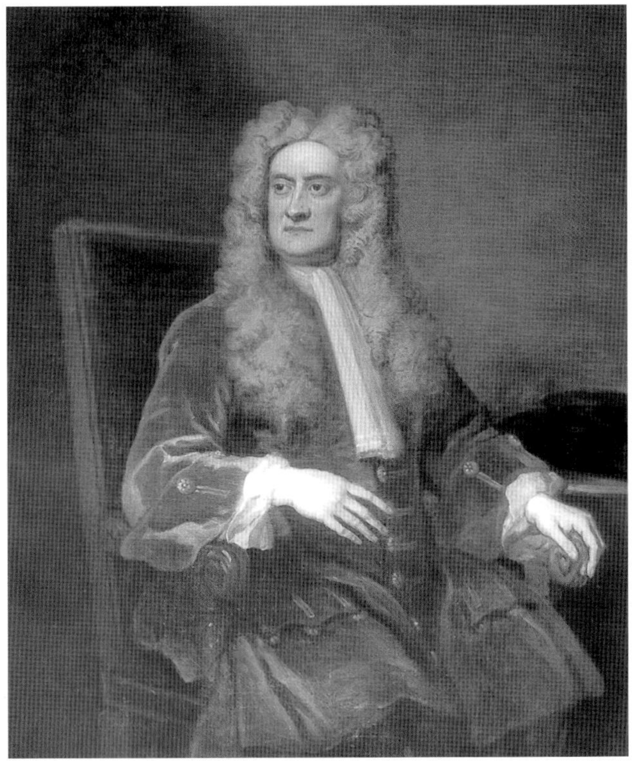

VIII-1
see colour Plate V

Three-quarter-length, seated in high-backed chair, with head half-turned to left, and eyes looking left; in full grey wig; body turned half-right, wearing buttoned green coat, with tied white cravat and white shirt cuffs; right elbow resting on arm of chair with hand lying across abdomen, and left hand grasping pommel of chair arm; on right a ledge beside the chair, with a dark green, folded drape lying on it.

Signed and dated on low left:

GKneller f 1719 (GK in monogram)

Provenance: Sir Thomas Parker, FRS, 1st Earl of Macclesfield (1666?–1732), for Shirburn Castle, which he had purchased in 1716 and added to and rebuilt.

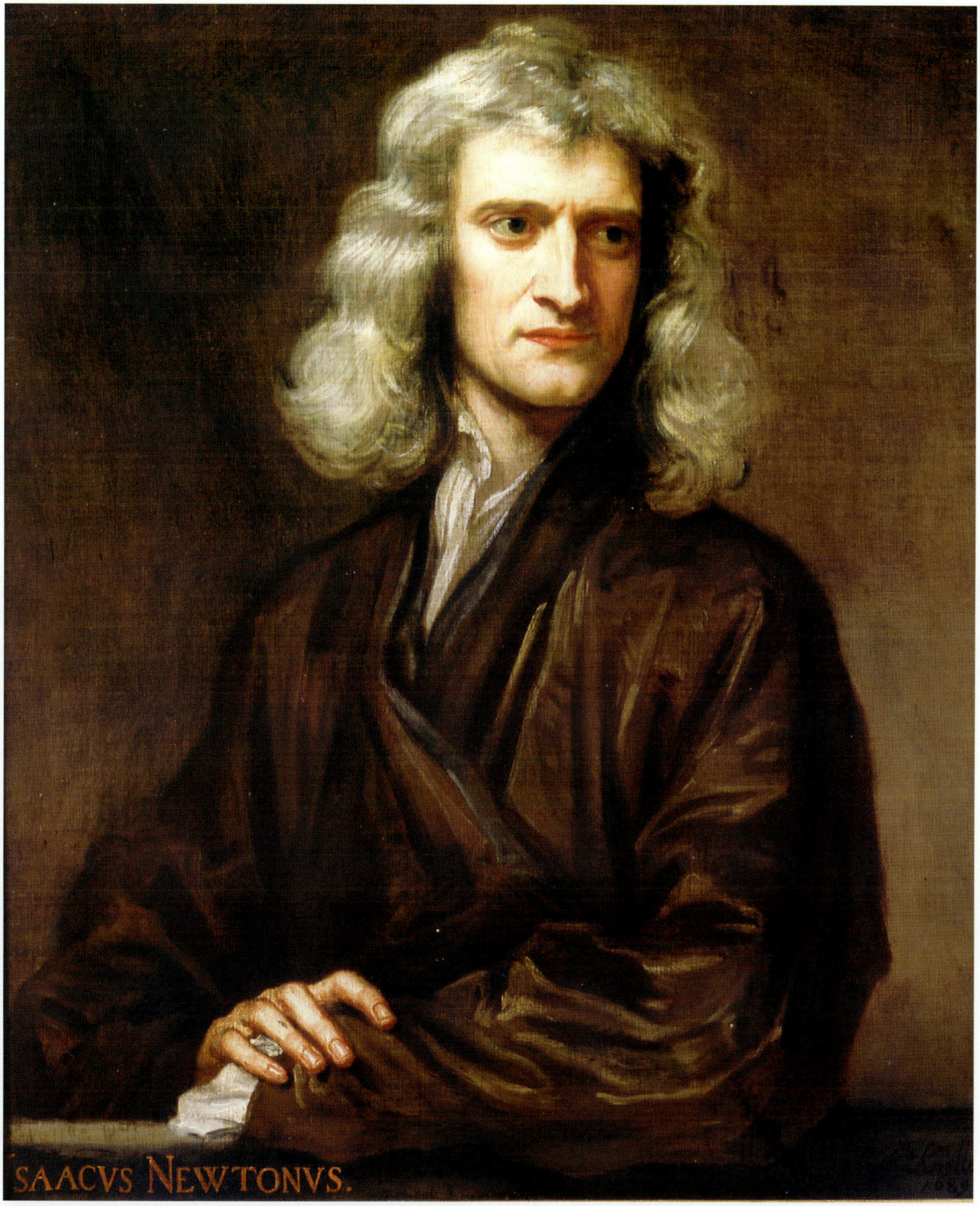

II-1 Newton aged forty-six, oil on canvas, by Sir Godfrey Kneller, 1689.

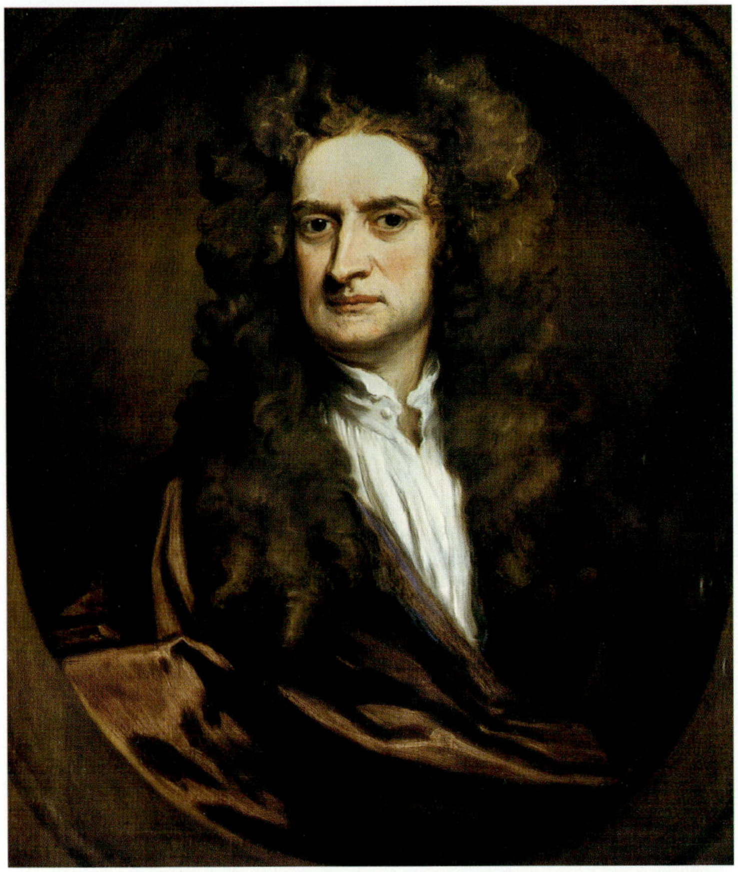

LEFT: **III-1** Newton aged fifty-nine, oil on canvas, by Sir Godfrey Kneller, 1702.

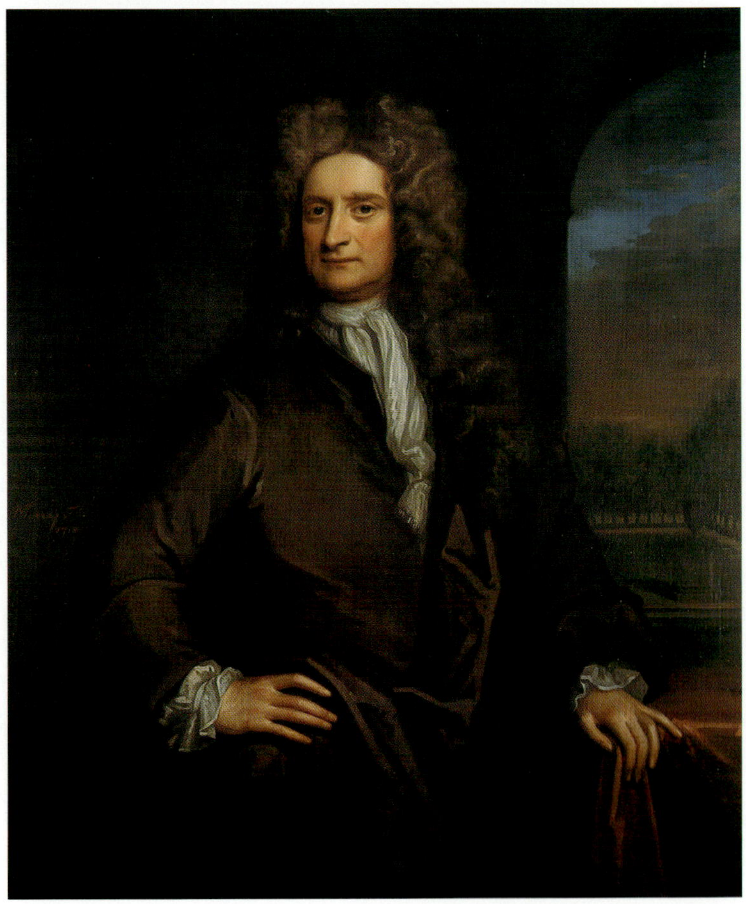

RIGHT: **VII-1** Newton aged seventy-five, oil on canvas, by Thomas Murray, 1718.

RIGHT: **IV–1** Newton aged sixty-eight, oil on canvas, by Sir James Thornhill, 1709–11.

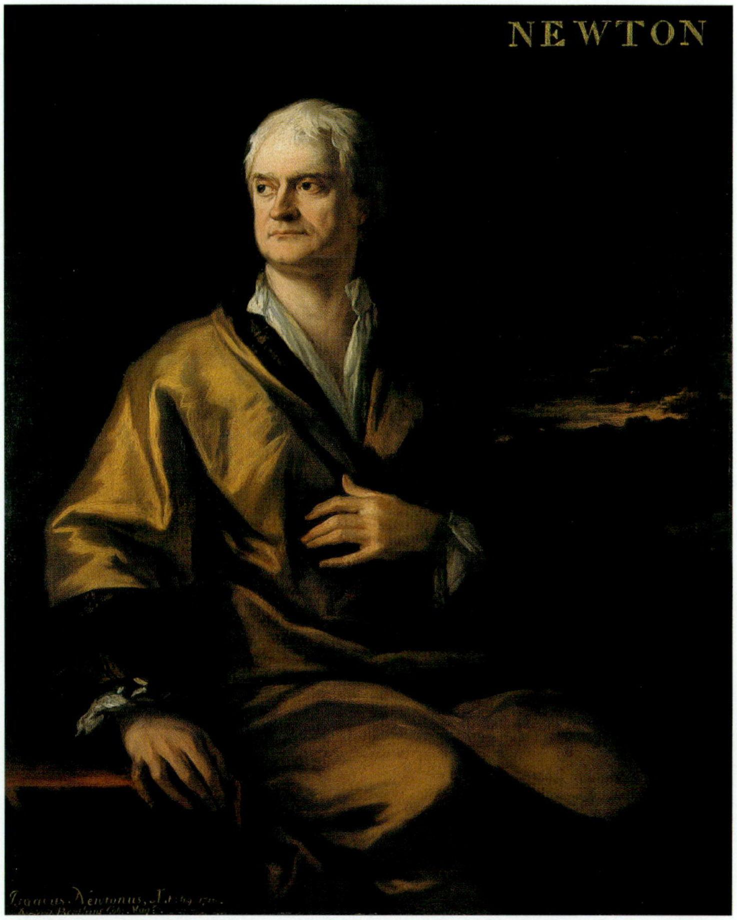

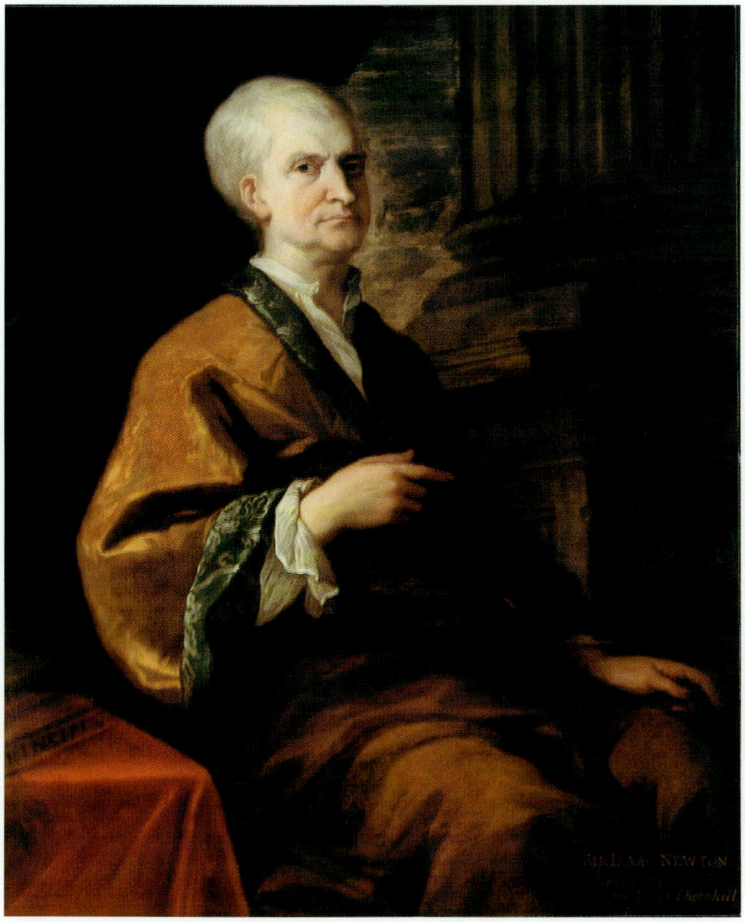

LEFT: **IV–2** Newton aged about seventy, oil on canvas, by Sir James Thornhill, c.1712.

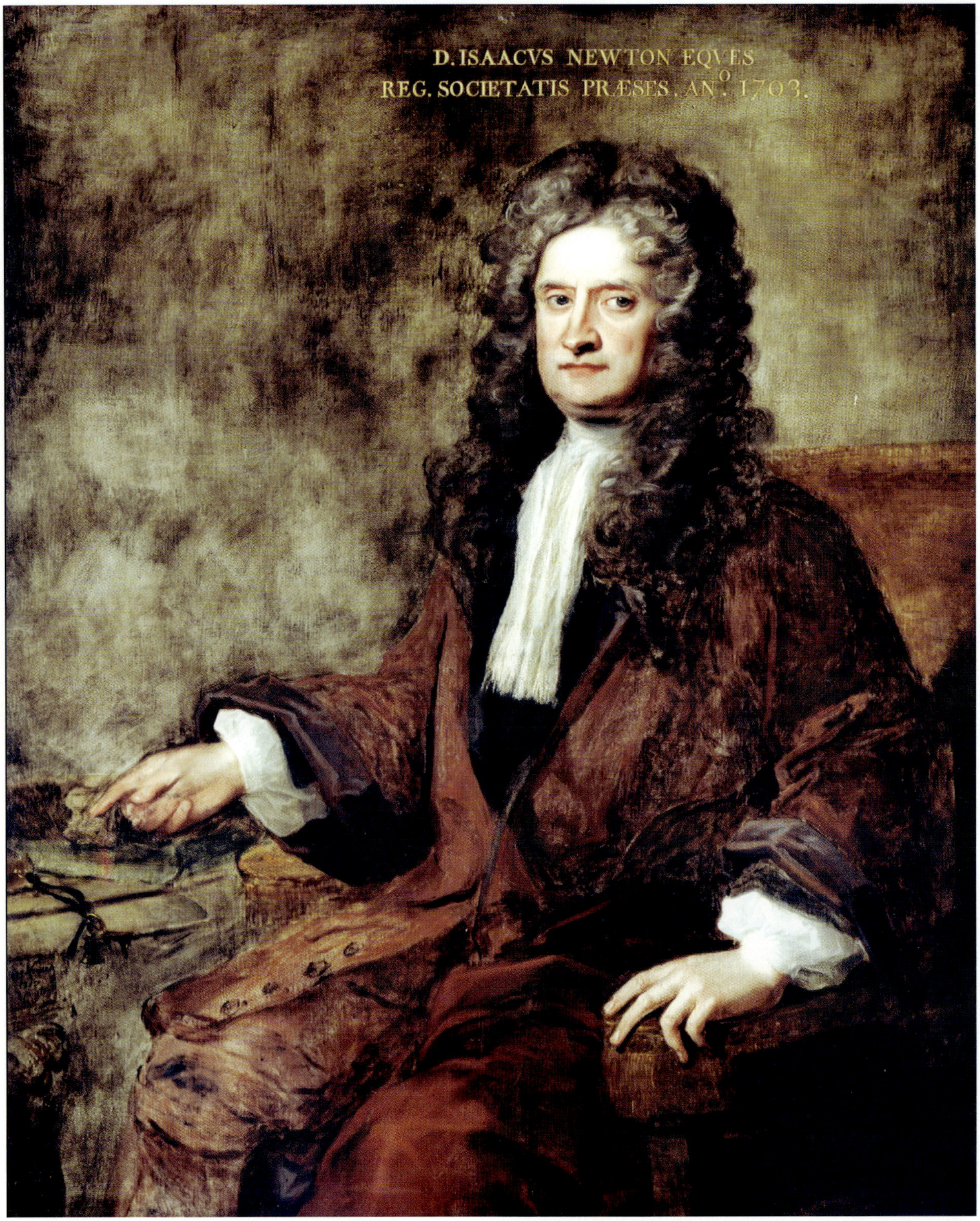

VI-1 Newton aged seventy-four, oil on canvas, by Charles Jervas, 1717.

RIGHT: **VIII-1** Newton aged seventy-six, oil on canvas, by Sir Godfrey Kneller, 1719.

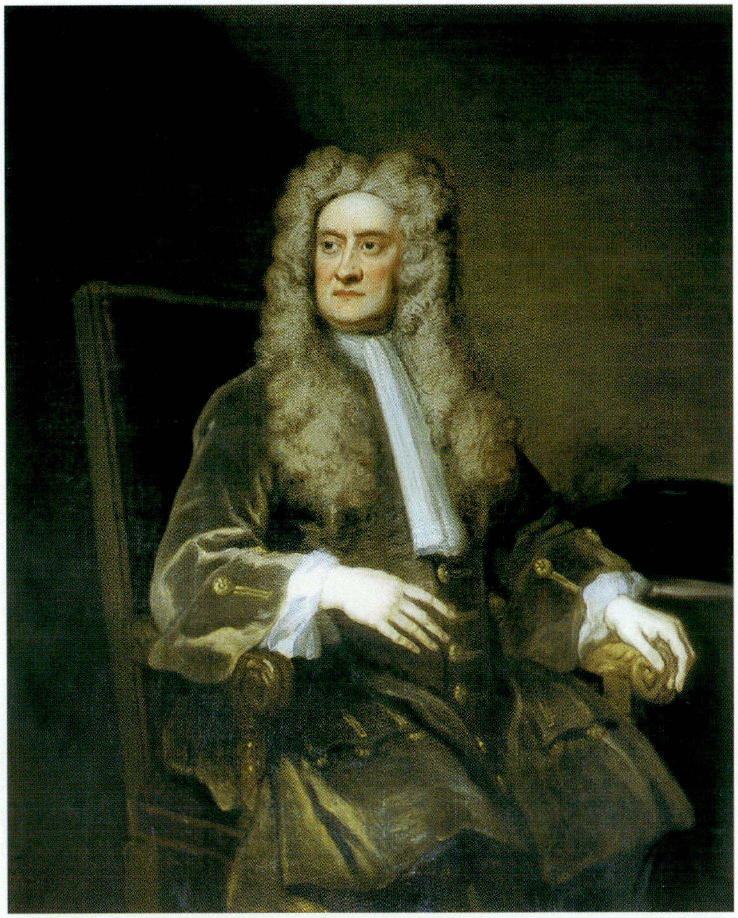

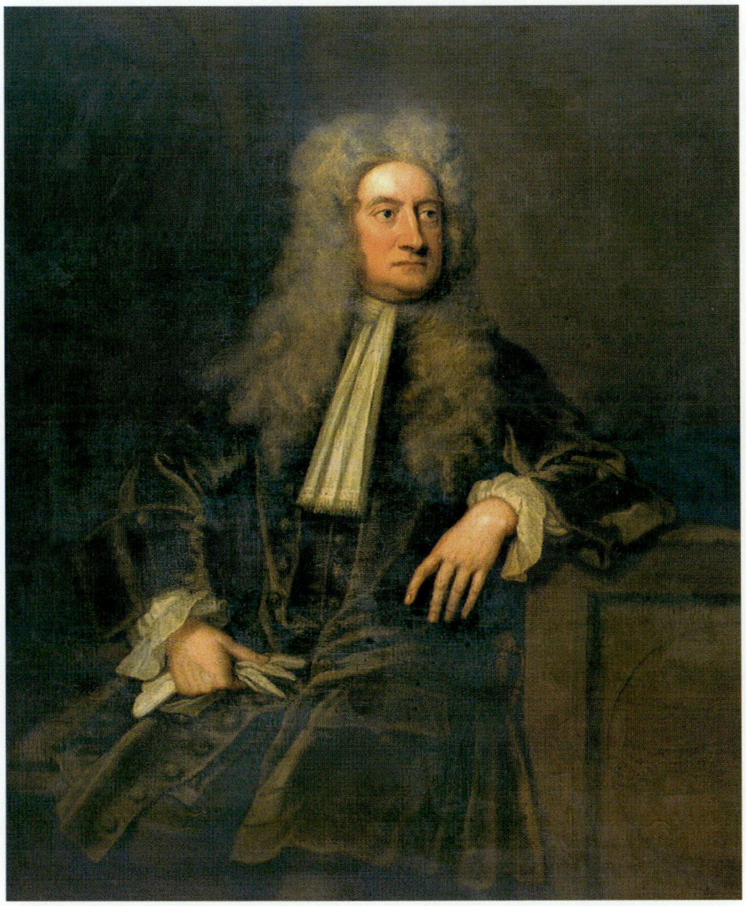

LEFT: **IX-1** Newton aged seventy-seven, oil on canvas, by Sir Godfrey Kneller, 1720.

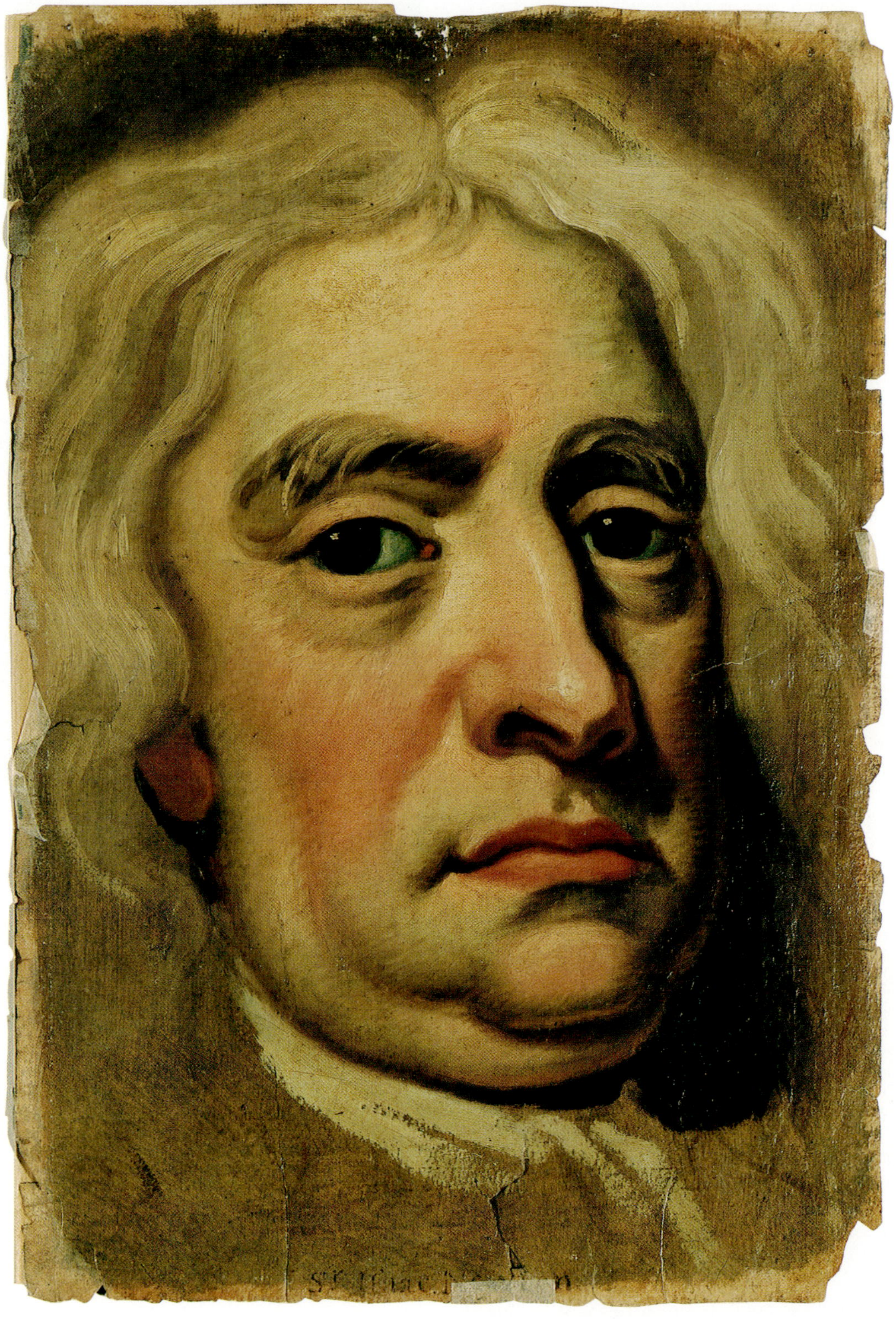

XIII-1 Head of Newton, oil on paper, by John Vanderbank.

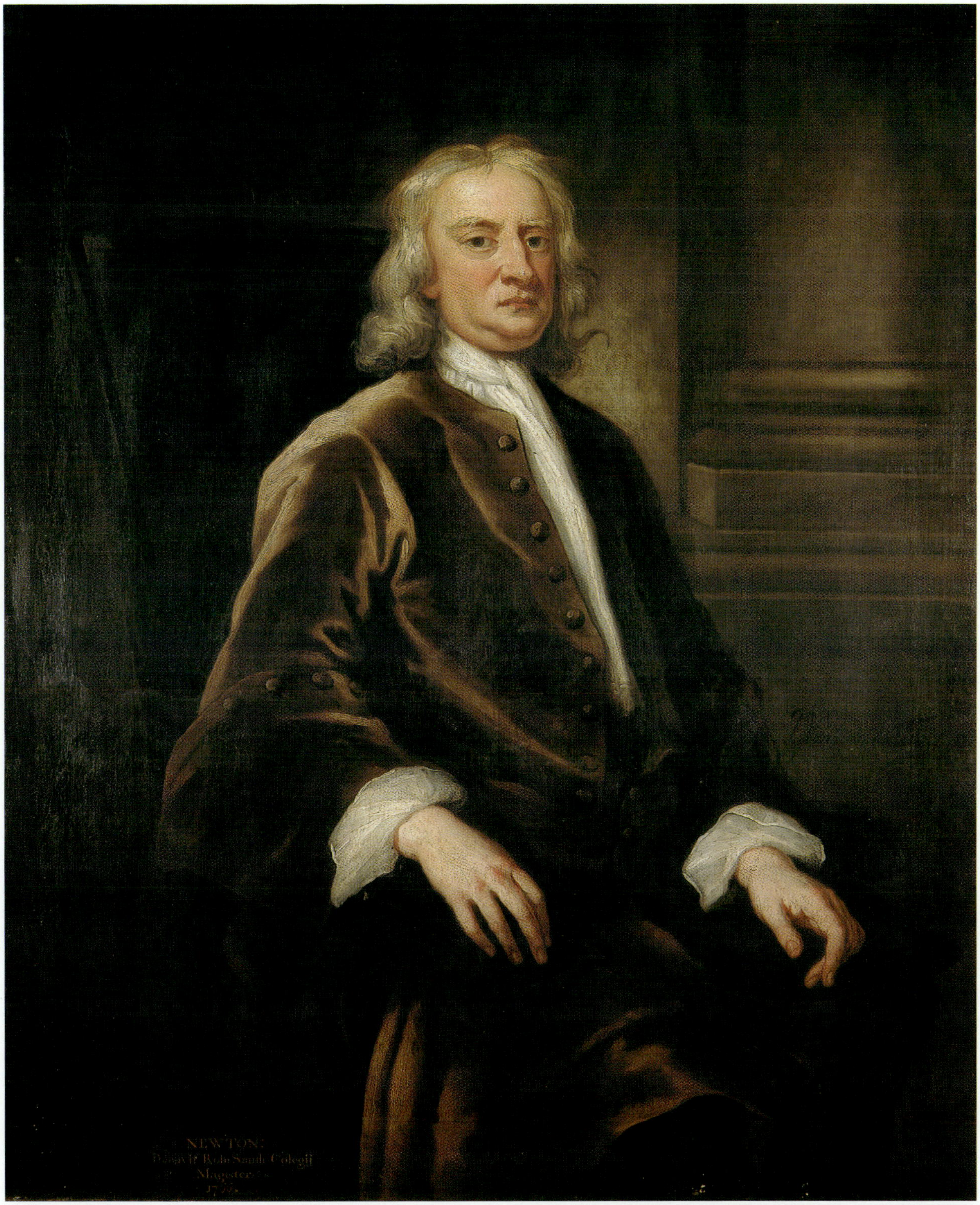

XIII-3 Newton aged eighty-three, oil on canvas, by John Vanderbank, 1726.

RIGHT: **XI-2** Profile portrait of Newton, pen and ink and wash on paper, by William Stukeley, 1726.

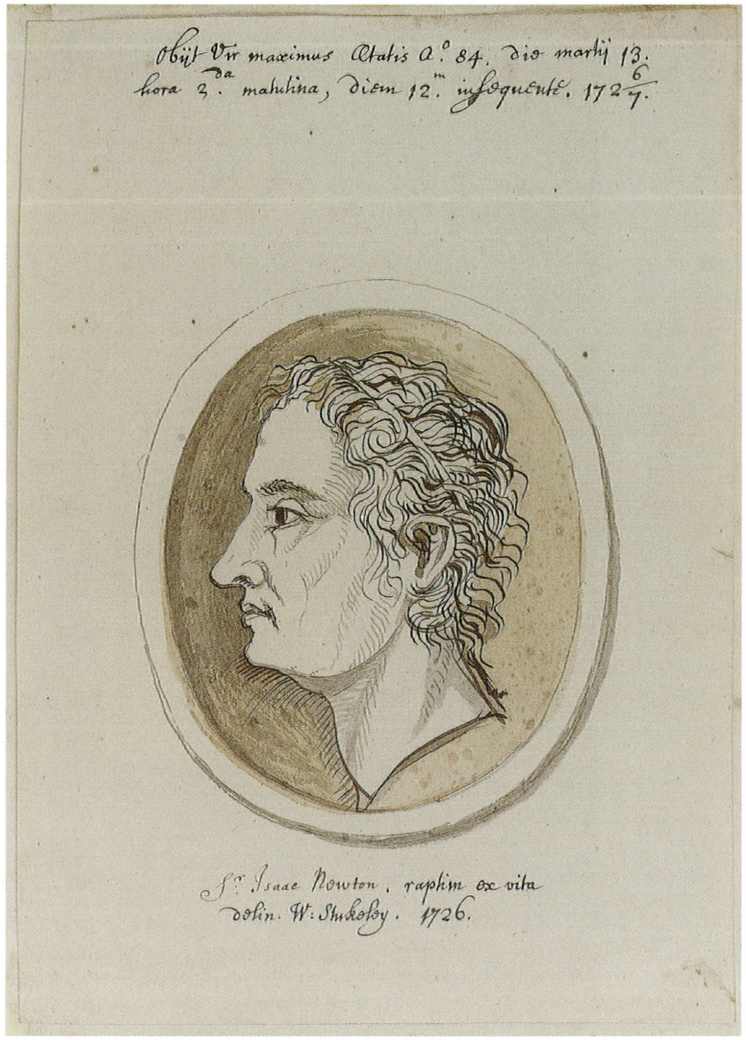

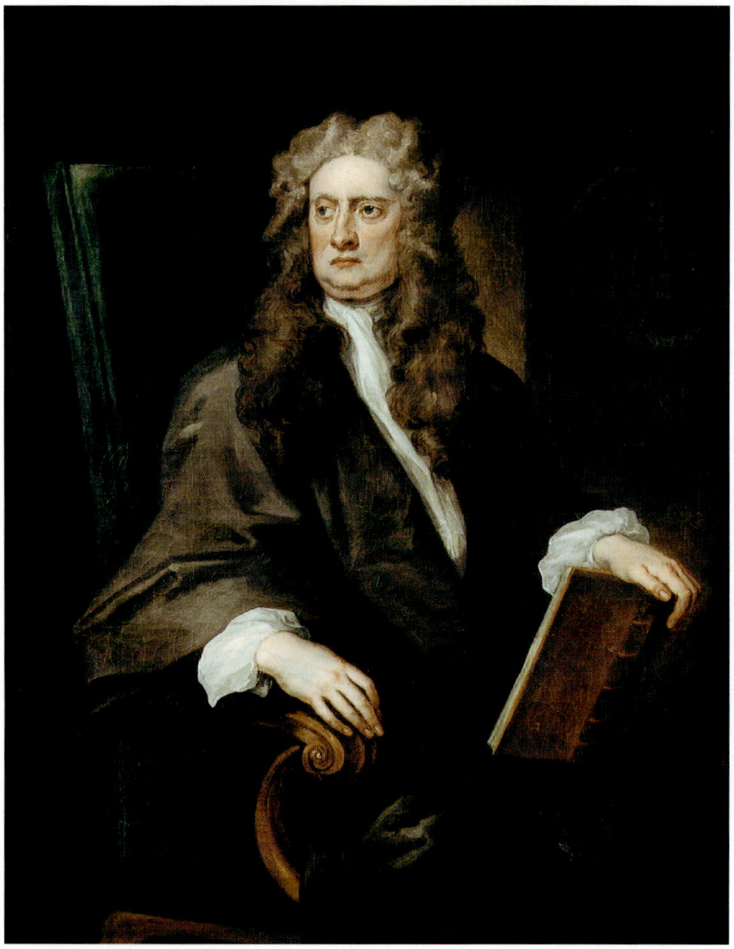

LEFT: **XV-1** Newton aged eighty-four, oil on canvas by John Vanderbank, 1727.

There is a companion, but unsigned, portrait by Kneller of Edmond Halley, FRS (1656–1742), three-quarter-length, wearing the uniform of a captain of the Royal Navy (not in Stewart[31]).

Stukeley[32] does not appear to have known of this portrait. He describes the painting of **IX-1**, dated a year later, and adds about it: 'There are a good many portraits of him [Newton] beside that by Sir Godfry Kneller abovementioned, which went into France. Sir Godfry painted two for Mr Conduit, one in his wig [**III-1**], as usual, another in his own hair [**II-1**]', both of which were kept by Newton till his death. This portrait was painted by Kneller when seventy-three years old.

VIII-2 Studio replica after **VIII-1**, oil on canvas.
127 × 100 cm (50 × 40 in)
Lane Fine Art, London, acquired in 2004 from France.

VIII-3 Jonathan Richardson, the elder (1665–1745)

Drawing after **VIII-1**, graphite on vellum.
13.4 × 11.9 cm ($6\frac{1}{8} \times 4\frac{5}{8}$ in)
British Museum since 1799 (GG-01-483).

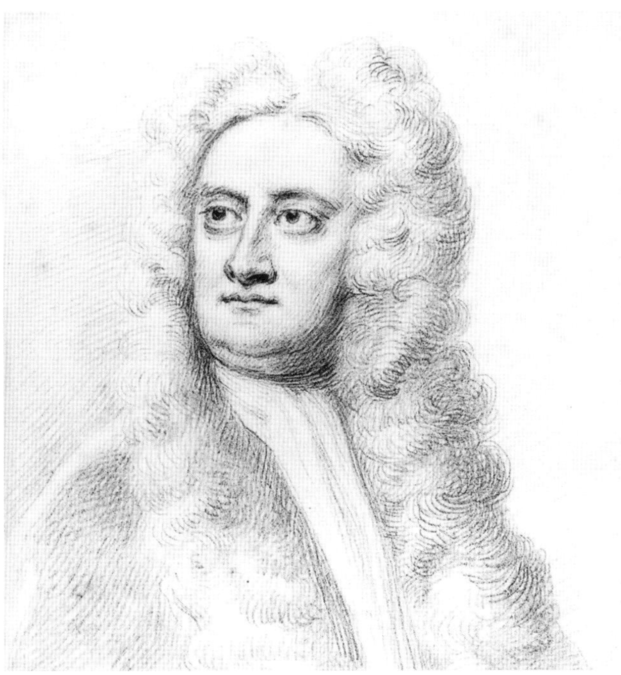

VIII-3

Half-length, seated figure, with head half-left, body slighly to right, wearing long curling wig and long cravat.

Collection: J. Richardson, the senior; J. Richardson (1694–1771), the younger; Revd C. M. Cracherod (1730–99).

IX. Sir Godfrey Kneller (1646–1723)

IX-1 Aged seventy-seven, oil on canvas, 1720.
122.5 × 99.7 cm ($48\frac{1}{4} \times 39\frac{1}{4}$ in)
Lord Egremont, Petworth House, West Sussex.
(Stewart[31] no. 530; Collins Baker[31] no. 153)

Three-quarter-length, seated figure, with head half-turned to right, and eyes looking forward; in full wig reaching to mid-chest; body turned slightly to left, wearing grey suit with pleated coat buttoned below waist, but exposing right knee in breeches, with a

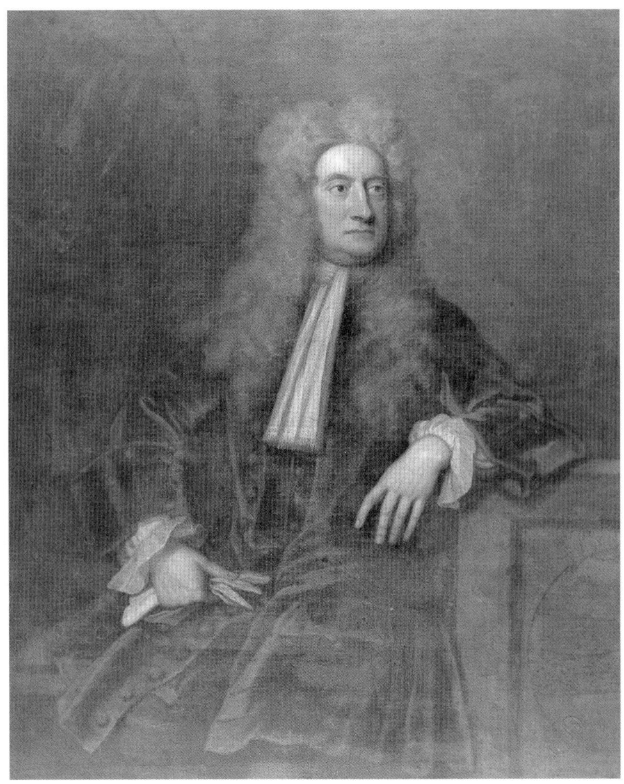

IX-1
see colour Plate V

sword-hilt between his right forearm and his waist; white cravat tied at throat, with equal ends reaching to lower chest, and white shirt cuffs; the right hand holding a pair of gloves is on his lap; the left arm is supported on a parapet, on which a sphere is marked; in left upper background is a recessed niche.

Signed and dated on back:

G. KNELLER BARONT. 1720.

On the appearance of the second edition of *Principia* in 1713, Newton sent copies to Abbé Jean-Paul Bignon (1662–1743), and to Bernard le Bovier de Fontenelle (1657–1757) and Pierre Varignon (1654–1722), for whom, as well, he also proposed membership of the Royal Society and sent copies of *Opticks*.[26] This was done to encourage these members of the Académie Royale des Sciences in Paris, of which he had been an associate member since 1699, to support him in his priority dispute with Gottfried Wilhelm Leibniz, FRS (1661–1733). In 1715, Varignon tried to procure a portrait of Newton, but with no success, until Newton commissioned this one for him in 1720, and also expressed a desire to be given a portrait of Varignon in return, which he received in January 1721.

In 1752, in his *Memoirs of Sir Isaac Newton's Life*,[32] William Stukeley, FRS (1687–1765) wrote:

> In the year 1720 Sir Isaacs picture was painted by Sir Godfry Kneller to be sent to Abbé Bignon [whose name he wrote in misrecollection for that of Varignon] in France, who sent his picture to Sir Isaac. Both Sir Isaac and Sir Godfry desir'd me to be present all the times of sitting, and it was no little entertainment to me, to hear all the discourse that passed between these two great men. Tho' it was Sir Isaac's temper to say little, yet it was one of Sir Godfrys arts to keep up a perpetual discourse, to preserve the lines and spirit of a face. I was delighted to observe Sir Godfry, who was not famous for sentiments of religion, sifting Sir Isaac to find out his notions on that head, who answered him with his usual modesty and caution.

After receiving the portrait and sending Newton his own in return, Varignon wrote:

> Dans ce portrait de M. Newton, il ne paroist pas avoir plus de 50 ans, tant il y a l'air frais & vigoureux: cependant M. Taylor [Brook Taylor, FRS (1685–1731)] ... me dist qu'il etoit fort ressemblant.[27]

Sir David Piper (1918–1990)[29] wondered whether, and Stewart[31] considered that, this signed and dated portrait (**IX-1**) was an autograph replica, if not the one painted for Varignon and sent to Paris. It is not known from where, and whether, it was acquired either by Charles Seymour, 6th Duke of Somerset (1662–1748), or by the 2nd Earl of Egremont (1710–1763), or, perhaps most likely, by the 3rd Earl (1751–1837); it is not in the 1763 inventory for Petworth or Egremont House, but is in a catalogue of 1835 at Petworth when it was hanging in the Ante-Room of the Carved Dining Room with woodwork and painting surrounds by Grinling Gibbons (1648–1720), and in an inventory of 1837 among forty-three paintings attributed to Kneller at Petworth. Kneller became a baronet in 1715.

IX-2 Studio replica after **IX-1**, oil on canvas, 1721. 125 × 100 cm ($49\frac{1}{4} \times 39\frac{1}{2}$ in)
Grantham Museum (Lincolnshire County Council) since 1987.

Three-quarter-length, seated wearing a dark grey coat and holding a white glove in his right hand; left arm leaning on a parapet; wearing a white cravat and sword; recessed shell niche in upper left background.

Signed on canvas, lower left:

GKneller f. 1721 (GK in monogram)

Provenance: Mrs R. G. Warren; Christie's 24 July 1980 (lot 118); Sotheby's 12 March 1986 (lot 37); Lane Fine Art 1987.

X. Unknown artist

X-1 Portrait, oil on canvas, *c.*1715–20.
124 × 99.7 cm ($48\frac{3}{4} \times 39\frac{1}{4}$ in)
Professor A. Rupert Hall, Tackley, near Oxford, from Sotheby's in May 1985.

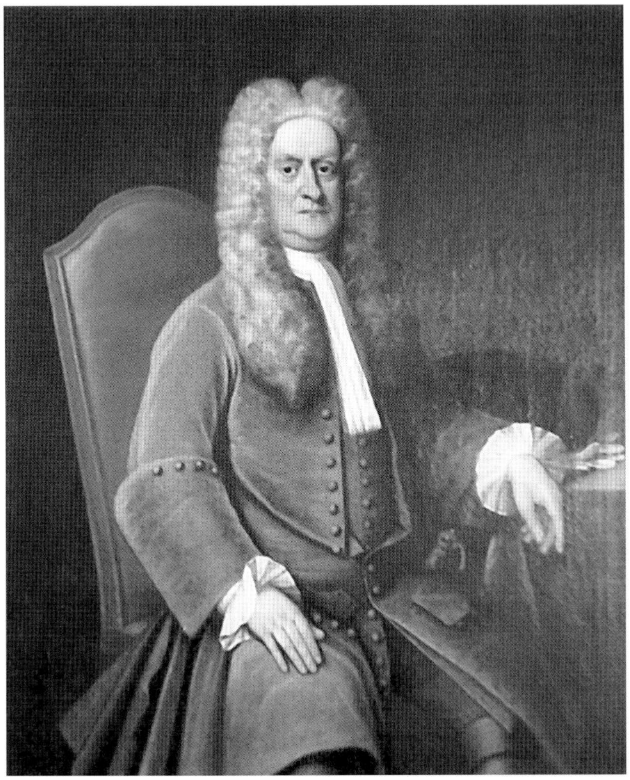

X-1

Three-quarter-length figure seated in tall-backed chair, with head turned slighly to his left but eyes looking forward, in full wig reaching to mid-chest; body half-turned to right in grey suit with pleated coat and tall cuffs; white cravat, and sword-hilt on left; right hand on right thigh, left forearm resting on table with pair of white gloves on it; dark background.

XI. William Stukeley, FRS (1687–1765)

XI-1 Aged about seventy-seven, drawing in pencil with yellow highlights on a page from a notebook.
19 × 15 cm ($7\frac{1}{2} \times 5\frac{7}{8}$ in)
Royal Society, London.

The many-breasted figure of Diana, goddess of light and representing the moon, is shown holding a rounded plaque (? the sun) that has Newton's head in profile on it. She is seated on a celestial globe, on which are drawn the seven stars of the plough, referring to the Argonauts and thus symbolising Jason's fire-breathing oxen plough. Above, two comets are added, which, besides providing a reference to prophecy, chronology and the *Principia*, were for Stukeley an important association with Newton. Newton in *The chronology of ancient kingdoms amended.* (1728, pp. 86–96) discusses observational data of various stars recorded in 1689–90, that he claims allow the dating of

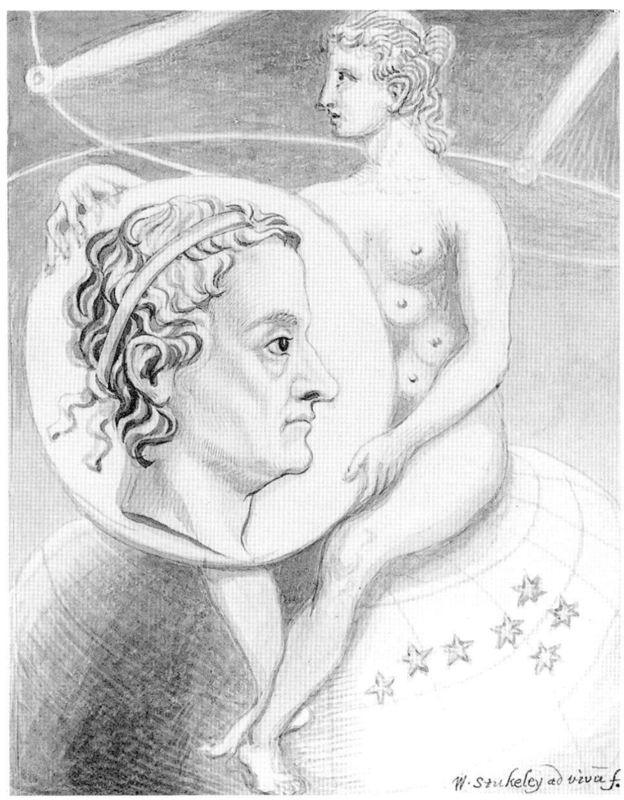

XI-1

the argonautic expedition from the astronomical evidence in its written record. He calculated (p. 95) that the voyage of the Argonauts took place in 937 BC, about forty-three years after the death of Solomon, a key date in Stukeley's own ancient chronology.

At lower right corner:

W. Stukeley ad vivu f.

The drawing is part of the manuscript of *Memoirs of Sir Isaac Newton's Life* by William Stukeley, MD, FRS, 1752, sold by his great-great-great-grandson, Oliver S. F. St. John, in Sotheby's, 1931, now in the Royal Society Archives, MS 142, and published[32] in 1936. In it Stukeley wrote that, over the years from 1717 until 1726 when he moved from London to Grantham, he made several drawings of Newton 'from the life, chiefly in the antique way of profile [i.e. as a Roman emperor]; and very like; whereof that here prefixed is one'. He added: 'I painted his profile, as big as life, in a niche in one of the wings of the garden front of my house at Grantham … underneath inscribed GENIO LOCI.' In a letter to John Conduitt on 22 July 1727, Stukeley wrote how he had, when he was present with Newton and Kneller:

> at painting his great picture [in 1720 – see **IX-1**] to be sent to France, desired Sir Isaac to let Sir Godfrey paint his side face, a profile as we call it, for me. 'What!' said Sir Isaac, 'Would you make a medal of me?' and refused it though I was then in the highest favour with him.[6]

Stukeley's drawings of Newton now seem mainly lost – there is none of Newton among the large collection of Stukeley's sketches in the Bodleian Museum. An epigram by Michael Maittaire (1668–1747), a classicist and typographical historian at Oxford, reads in reference to one of them:

Newtoni expressit vultum tua linea; mentem
Expressisse viri linea nulla potest.[32]
[Your art has produced a likeness of Newton's face,
but no art has the power to show his mind.]

A profile medallion portrait of Maittaire, in pen and ochre washes and dated 15 December 1721, by Stukeley was included with **XI-2** and **XI-3** at Bonhams, 18 December 2002 (lot 788).

XI-2 Profile medallion portrait, facing left, within an oval, drawn on paper in pen-and-ink and ochre and grey washes and pasted into an album, 1726.
25.7 × 18 cm ($10\frac{1}{8} \times 7\frac{1}{8}$ in)
Bonhams, 18 December 2002 (lot 788).

Inscribed at top of page (with the day of Newton's death incorrectly):

obiit vir maximus AEtatis A^{o} *84. die martii 13.*
hora 2da*. matutina, diem 12m. insequente. 1726/7.*
[The greatest man died at the age of 84 on the 13th day of March at the second hour of the morning, the day following 12 March 1726/7]

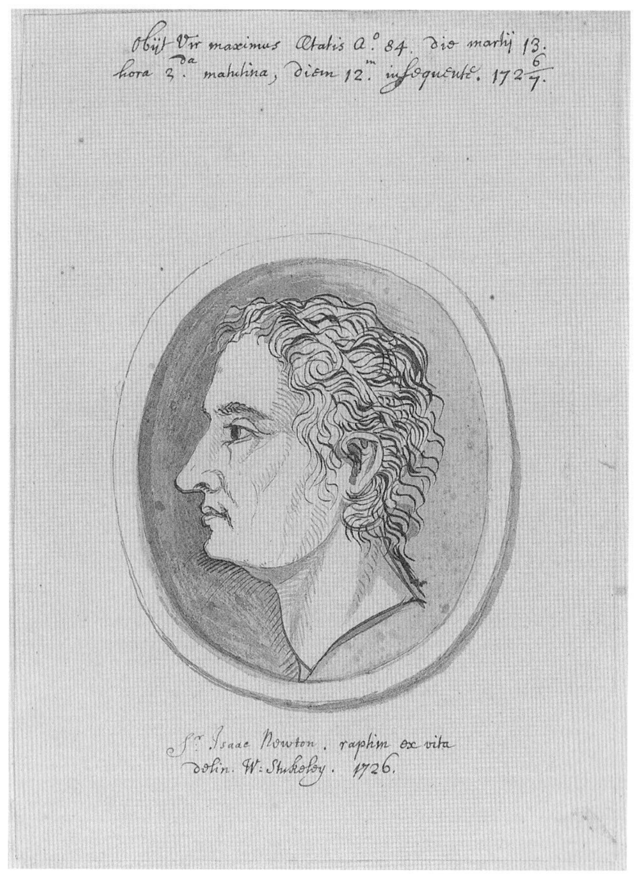

XI-2
see colour Plate VIII

Inscribed below drawing:

S^r: Isaac Newton, raptim ex vita
delin. W: Stukeley. 1726.
[Sir Isaac Newton, drawn hurriedly from life.
W. Stukeley. 1726]

XI-3. Lightly sketched profile medallion portrait, facing left, within an oval inscribed beneath 'NEUTONVS', in grey washes, on trimmed page, with traces of inscription below, pasted into an album.
17 × 14.5 cm ($6\frac{5}{8} \times 5\frac{3}{4}$ in)
Bonhams, 18 December 2002 (lot 788).

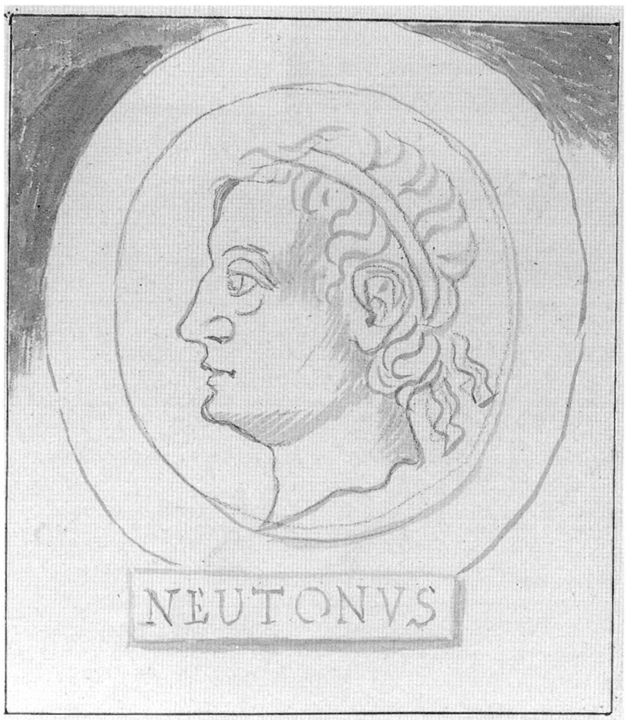

XI-3

XII. Michael Dahl (1656–1743)[8]

XII-1 Head of Newton 'in his hair', that is, without a wig, by 1724.
Second Earl of Oxford and Mortimer (1689–1741) at Wimpole Hall, Cambridgeshire. The portrait has been missing since 1742.

Mentioned by George Vertue (1684–1756) in his *Notebooks*[33] in 1724 and in 1733 as hanging in the new Great Library Room, built by James Gibbs (1682–1754) and thought to have been completed in 1720, at Wimpole Hall. The Duke of Newcastle (1662–1711) bought Wimpole Hall in 1710, but died in a riding accident the following year, leaving his only child, Henrietta Cavendish-Holles, as his heiress. She married Edward, Lord Harley, the son of the politician, Robert Harley, 1st Earl of Oxford (1661–1724), at Wimpole in 1713, and he, as well as adding to his father's books, prints and manuscripts, made a large collection of paintings. Because of a financial crisis, he sold Wimpole, to Philip Yorke (1690–1764), later 1st Earl of Hardwicke, in 1740, but died in 1741. Lady Oxford sold the collection of sculptures and three hundred old master paintings and portraits of famous persons over six days from 8 March 1742,[8] and the Dahl portrait of Newton was sold on 12 March.

XIII. John Vanderbank (Johan van der Banck) (?1694–1739)

XIII-1 Head of Newton, study in oil on paper.
30.5 × 20.3 cm (12 × 8 in)
Royal Society, purchased in 1950.

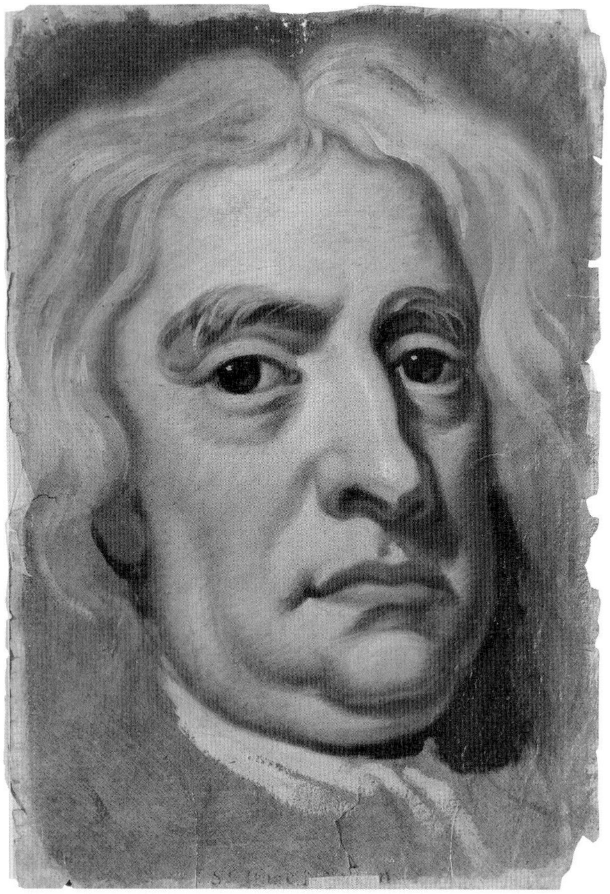

XIII-1
see colour Plate VI

This is the study for portraits **XIII-2** and **XIII-3**.

XIII-2 Aged eighty-three, oil on canvas, 1726.
127 × 101.6 cm (50 × 40 in)
Royal Society, London.

Three-quarter-length, life-size seated figure, turned and facing half-right, looking front; bare-headed, long, rough, white hair; seated in tall-backed chair, with right hand on its arm and left hand on thigh; wears collarless

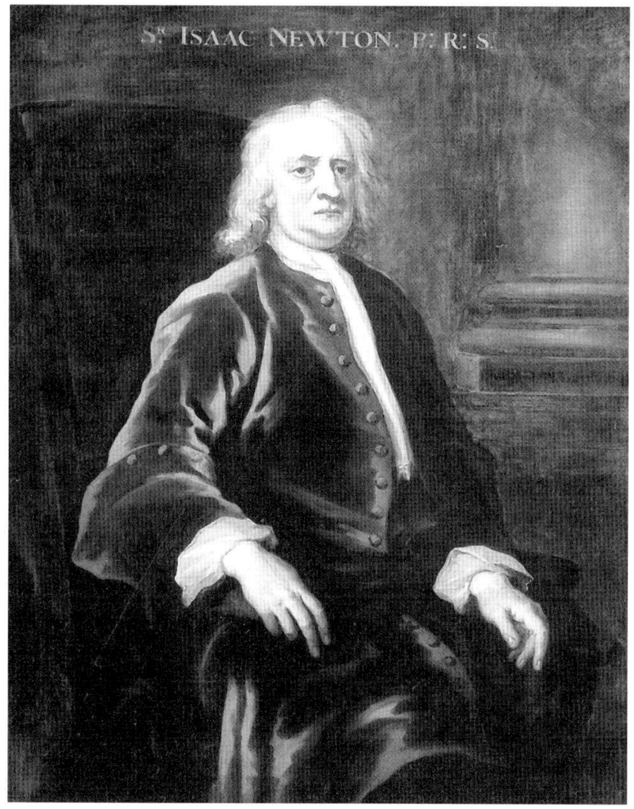

XIII-2

coat, buttoned at waist, and soft white scarf round neck, tied below chin, with ends reaching to waist under coat; white shirt cuffs turned back over coat sleeves. Base of pillar in right background.

On canvas, across top:
S^{R} ISAAC NEWTON. P: R: S.
on base of pillar:
I NEWTON EQ. R.S.PR.
to right of left forearm:
J. Vanderbank Fecit
1725.

Presented to Royal Society by Charles Blacker Vignoles, FRS (1793–1875) in 1841. The first of eight engravings of this portrait, a mezzotint (**XLVII-1**), has the inscription 'J. Faber fecit 1726. J. Vanderbank pinxit 1725. Isaacus Newton Equ. Anno Aetat+ 84 A.D. 1725. J. Cooper excudit.' Newton entered his eighty-fourth year on Christmas Day, 1725, which implies the portrait was painted between then and 25 March 1726.

XIII-3 Aged eighty-three, another version of **XIII-2**, oil on canvas, 1726.
127 × 101.6 cm (50 × 40 in) Trinity College, Cambridge.

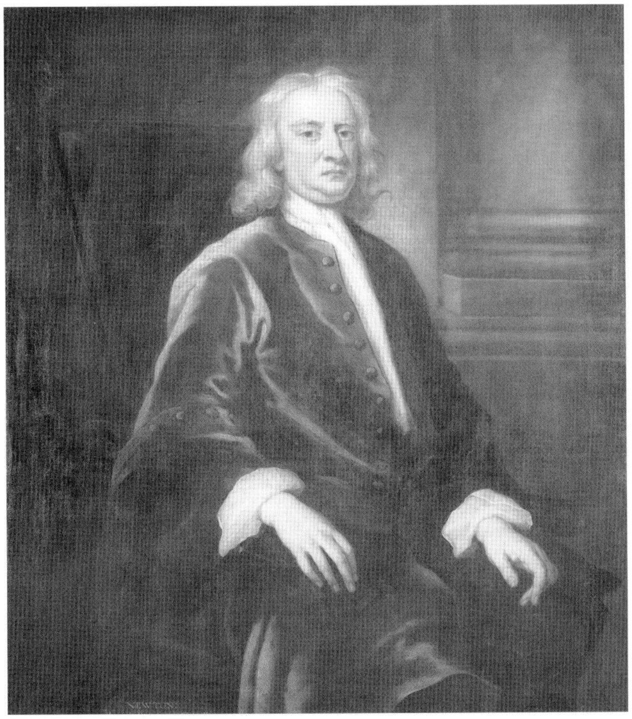

XIII-3
see colour Plate VII

Three-quarter-length, life-size seated figure; head and body half-right, with long, wavy hair falling to shoulders; seated in high-backed chair, with right hand on its arm and left hand on thigh; wears brown collarless coat, buttoned at waist, and soft white scarf round neck, tied under chin, with ends reaching to waist under coat; white shirt cuffs turned back over coat sleeves. There is the base of a pillar in right background.

On canvas, on base of pillar:

ISAAC NEWTON. EQ.

to right of left forearm:

J. Vanderbank Fecit
1725.

and low left:

NEWTON./Donavit Rob. Smith
Collegij/Magister/1766.

Robert Smith, FRS (1689–1768) was Master of Trinity College 1742-68. The painting of the right edge of the coat, the cravat and the hands differs from **XIII-2**.

XIII-4 George Vertue (1684–1756)

Bodycolour on vellum after **XIII-2**, 1726.
19 × 15.6 cm ($7\frac{1}{2} \times 6\frac{1}{8}$ in)
Private collector, New York, after Christie's, 21 March 1989 (lot 73).

Life-size seated figure; head and body half-right with long wavy hair falling to shoulders; seated in high-backed chair, right hand on arm of chair, left hand on knee; wears collarless coat, unbuttoned above and below waist, showing soft white scarf round neck, tied under chin with ends reaching to waist under coat; white cuffs turned back over coat sleeves. Base of pillar in right background.

Signed in full and with monogram, and dated 1726.

The portrait indicates that Newton, who was a short man (5 feet 6 inches tall), had become plump by the age of eighty-three. Engraved (**XLVIII-1**) by G. Vertue for frontispiece to *Philosophiae Naturalis Principia Mathematica.* 3rd edn 1726,[25] the first edition to have a portrait of Newton in it.

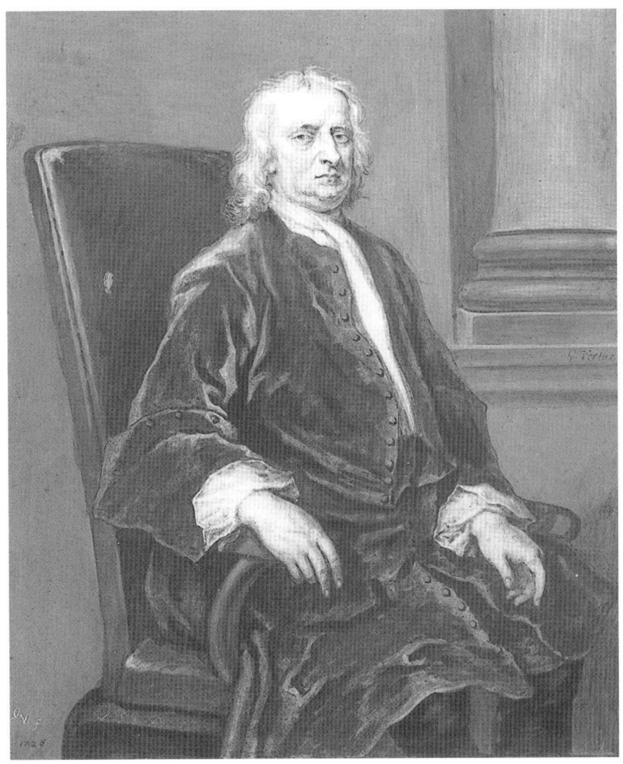

XIII-4

XIII-5 Bust-length, studio copy after **XIII-3**, oil on canvas.
Oval, 75.5 × 63.5 cm (29¾ × 25 in)
Royal Astronomical Society, Burlington House, London.

The cravat and the right edge of the coat with its buttons are similar to **XIII-2**.

XIII-6 Bust-length, studio copy after **XIII-3**, oil on canvas.
76 × 64 cm ($29\frac{15}{16}$ × $25\frac{3}{16}$ in)
Library, Royal Military Academy Sandhurst, Camberley, Surrey.

The painting is in poor condition with overpainting.

XIII-7 Bust-length, studio copy of **XIII-2**, oil on canvas.
76.2 × 63.5 cm (30 × 25 in)
California Institute of Technology, stolen in 1979 and since missing.

This portrait is differed in the tying and painting of the cravat and has a plain dark background.

It was at the Arthur U. Newton Galleries, New York in 1929, purchased from the Dalzell Hatfield Gallery, Los Angeles, by Ralph Lloyd in the 1930s, and subsequently donated to Caltech.

XIII-8 Jonathan Richardson, the elder (1665–1745)

Drawing in pencil after **XIII-1**, 1733.
14 × 12 cm (5½ × 4¾ in)
Collection: Roger Warner in 1955.
Inscribed on back:

Sir Is. Newton
3 Jan 173⅔

XIII-9 Contemporary copy after **XIII-2**, oil on canvas.
127 × 101.6 cm (50 × 40 in)
Old Schools, University of Cambridge.

Collection: Revd T. A. Walker (1862–1935), sold 1924; W. Heffer & Sons, Ltd, Cambridge, from whom it was bought by the Fitzwilliam Museum (no. 1136 – 1924); deposited in Old Schools in 1935.[16] An indifferent eighteenth-century copy.

XIV. Enoch Seeman (1694–1744)

XIV-1 Aged eighty-three, oil on canvas, 1726.
43.2 × 33 cm (17 × 13 in)
Trinity College, Cambridge.

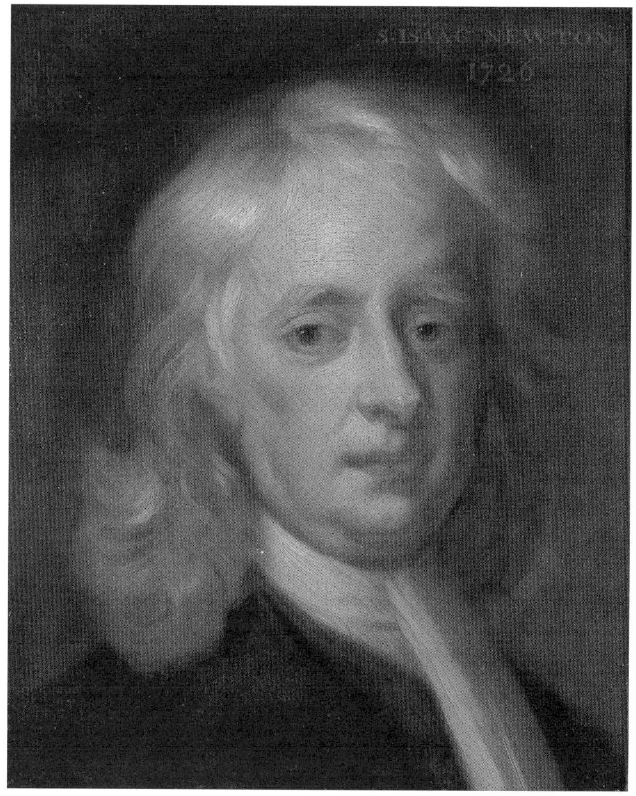

XIV-1
see colour Plate IX

Bust-length, nearly life-size; head half-turned to right; long grey-white hair, light brown eyes to front, face shaven; shoulders half-right; wears dark coat and white cravat over shirt.

On canvas high right:

Isaac Newton 1726

On back:

Newton painted by Enoch Seeman given by Holles – see his Memoirs Vol II

This study for portrait **XIV-2** was given to Trinity College in 1761 by the antiquary Thomas Hollis, FRS (1720–1774).[18] 'Mezzotint print of Sir Isaac Newton drawn and scraped MDCCLX by James Macardel from an original portrait painted by Enoch Seeman now in possession of Thomas Hollis F.R. and A.S.S.' appears in *Memoirs of Thomas Hollis*, vol. II, p. 800, 1780[18] – see **XLIX-1**.

XIV-2 Aged eighty-three, oil on canvas, 1726.
124.5 × 127.7 cm (49 × 50¼ in)
Justin Coldwell, Linley Hall, near Bishop's Castle, Shropshire.

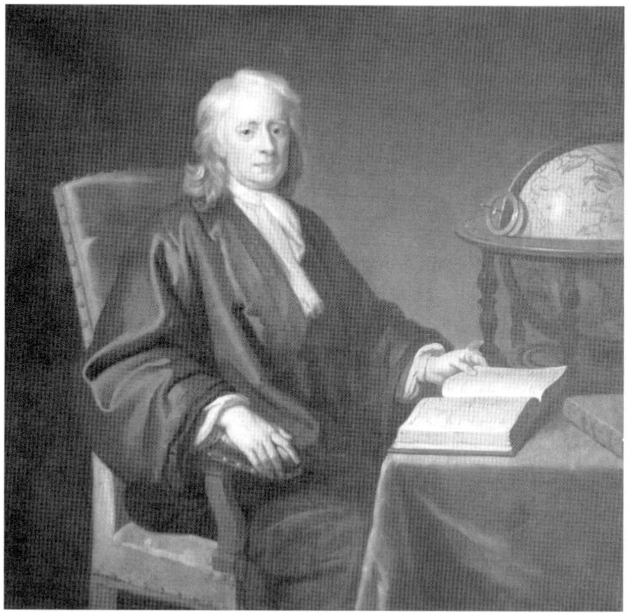

XIV-2
see colour Plate IX

Three-quarter length, plump seated figure; body three-quarter facing right, head slightly turned to his right; grey-white curly hair falling over shoulder, dark grey-blue eyes looking forward; right arm resting on the arm of chair with gap between index and middle fingers, left hand holding the pages of an open book lying on a table on the right; wears a blue gown, with a loosely tied white neckcloth covering a white shirt with cuffs; seated in high-backed chair at a table covered by a red tablecloth; on the table beside the open book is a celestial globe and in front of it is a closed and untitled book.

Provenance: By descent from Robert More, FRS (1703–1780), who replaced an earlier house with the present one in 1742–5. More's portrait, painted by Enoch Seeman in the late 1720s, hangs near **XIV-2**.

In this painting, **XIV-1** is closely followed. On the right-hand page of the open book there is a diagram taken from p. 299 of the third edition of *Principia*,[25] published in 1726. The third edition was the first to have a frontispiece, an engraving by George Vertue made from his portrait of Newton (**XIII-4**) after the Vanderbank (**XIII-2**) of 1726 at the Royal Society. The portrait at Linley Hall is framed in a similar manner to four other eighteenth-century portraits also hanging in the hall on the first floor.

XIV-3 Aged eighty-three, after **XIV-2**, oil on canvas.
125.7 × 100 cm (49½ × 39⅗ in)
Trinity College, Cambridge.

Three-quarter length, seated figure; body three-quarter facing right, head slightly turned to his right, with long white hair falling over shoulders; right hand on arm of chair with gap between index and middle fingers, left hand holding a book, the *Principia*, of which both of the two open pages show a diagram; wears plain blue gown over brown coat with buttons, and loosely tied white

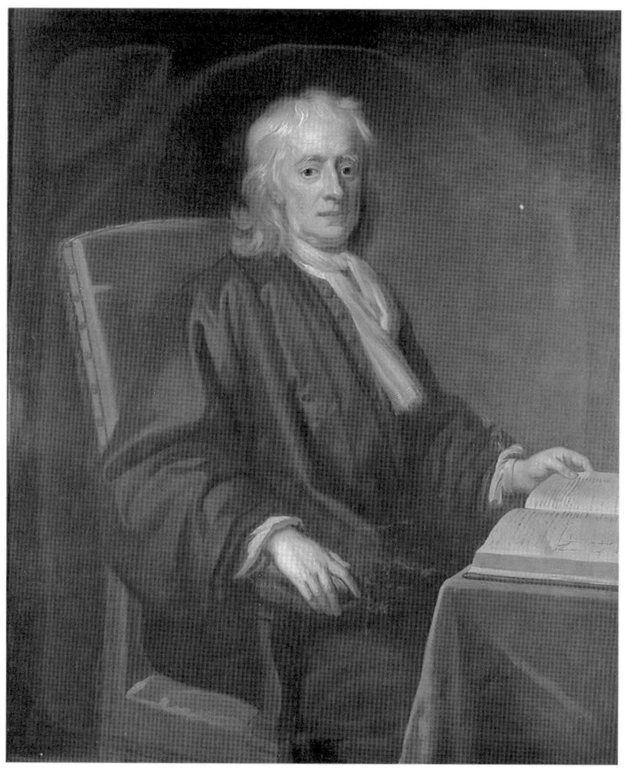

XIV-3
see colour Plate X

neckcloth, with long ends. Seated in high-backed chair, upholstered in red, before a table with red tablecloth.

The handling is close to **XIV-1** and to **XIV-2**. The diagram on the left-hand open page is taken from p. 256, and the one on the right-hand page from p. 257, of the third edition of *Principia*.

The portrait was bought by the College in 1850 from Mr White, dealer of 14 Brownlow Street, Holborn, who had purchased it from the executors of Lady Harley.[11] Lady Frances Harley, sister of the sixth and last Earl of Oxford, died in 1849. The purchase was advocated by Adam Sedgwick, FRS (1785–1873), Fellow of Trinity College, and appears in the Bursar's accounts for 1851: 'Portrait of Sir Isaac Newton by Vanderbank £105.' George Vertue (1684–1756) mentions in his *Notebooks*[33] of 1724 and 1733 a portrait of Newton 'in his hair' (**XII-1**) by Michael Dahl (1656–1743) at Wimpole Hall, Cambridgeshire, an estate which then belonged to 2nd Earl of Oxford but was sold in 1740 to Philip Yorke, later 1st Earl of Hardwicke. **XIV-3** could have been at Wimpole Hall from 1726 to 1740, before descent in the Harley family to 1849. However, stylistically it cannot be attributed to Michael Dahl, and in any case it cannot be Dahl's portrait of Newton, which was at Wimpole as early as 1724, and sold by Lady Oxford in 1742. The painting has often been attributed to Vanderbank since at least 1851.

XIV-4 Unknown artist, after Seeman **XIV-2** , oil on canvas.
127 × 148 cm (50 × 58¼ in)
Bequeathed by John Hatsell (1743–1820) to the British Museum in 1820, and transferred to the National Portrait Gallery (Piper[29] no. 558) in 1879.

Three-quarter length, seated figure; body three-quarter facing right, head slightly turned to his right, with long white hair falling over shoulders, and dark grey-blue eyes looking forward; right arm resting on the arm of chair with gap between index and middle fingers, left hand holding the pages of an open book; wears an ample dark blue robe with gold patterning over brown coat; plain white neckcloth loosely tied, with long ends; seated in high-backed chair at red-covered table on the right; on the table beside the open book is a celestial globe and in front of it a closed folio lettered OPTICKS on the spine; on the left, there are shelves filled with brown folios; in an oval carved on the side of the table, the crossed bones of the Newton family arms.

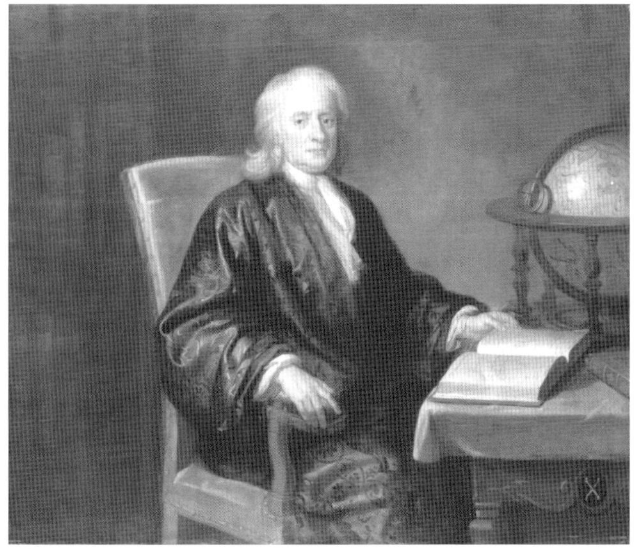

XIV-4
see colour Plate X

The book on the table shows three mathematical diagrams, one on the left hand and two on the right hand of its two open pages, that only appear on consecutive pages (pp. 204 and 205) in the third edition of the *Principia*,[25] published in 1726, and are different from the one diagram depicted in **XIV-2** and the two in **XIV-3**. The handling of the portrait, commonly attributed to Vanderbank in the past, cannot be his when compared with the portraits **XIII-1**, **XIII-2**, **XIII-3** and **XV-1** by him in the Royal Society and in Trinity College, nor is it in the hand of Michael Dahl. The Newton arms placed on the end of the table, and the other differences from **XIV-2**, such as the giving of its title to *Opticks*,[26] besides its larger size, imply that this portrait was made for the Newton family.

The painting relates to the Seeman portrait **XIV-2**, but less closely to **XIV-1**, **XIV-2**, **XIV-3** and **XIV-5** than they do to each other. It is apparently by a different hand, too, as its painting is less vigorous: this is shown in the way the artist has simplified and smoothed out Newton's features and hair. On the other hand, the three diagrams in it from the third edition of *Principia*, which differ from those in **XIV-2** and **XIV-3**, suggest that the painter of this portrait was Seeman, as another artist would have been unlikely to have shown such originality in making a copy, even for the Newton family. The celestial globe is very close to the globe in **XIV-2**.

XIV-5 Possibly by Seeman after **XIV-2**, oil on canvas.
124.5 × 99 cm (49 × 39 in)
Grace K. Babson Collection, Babson Institute, Wellesley, Massachusetts.[9]

This portrait was sold at Sotheby's on 26 June 1940 (lot 157). In left lower corner is written: *Sir Isaac Newton.* In the right background there is a celestial globe, not standing, however, on the table with the open book on it as in **XIV-2**: in the portrait **XIV-3** of the same size, to which it is otherwise close, the globe is not included, but here it has been squeezed in. The right-hand open page has a well-defined diagram on it, which does not appear in the third edition of *Principia*, and therefore differs from the diagrams in **XIV-2**, **XIV-3** and **XIV-4**.

XIV-6 Unknown artist, after **XIV-2**, oil on canvas.
76.2 × 63.5 cm (30 × 25 in)
American Philosophical Society, Philadelphia, purchased in 1947.

Bust-length, sitting figure, with body three-quarter right, head turned to his right, and eyes focussed on viewer; untidy shoulder-length white hair; white cravat tied at neck, with unbuttoned waistcoat.

XIV-7 Unknown artist, after **XIV-2**, oil on canvas within painted oval.
76.8 × 64.8 cm ($30\frac{1}{4} \times 25\frac{1}{2}$ in)
Los Angeles County Museum of Art; Sotheby's New York 29 May 2003 (lot 29A)

Bust-length, with body three-quarter right, head turned to right, and eyes focussed on viewer; untidy shoulder length hair, white cravat tied with lower end differs from **XIV-6**.

XIV-8 Unknown artist after **XIV-2**, watercolour on ivory.
Oval, 3.2 × 2.8 cm ($1\frac{1}{4} \times 1\frac{1}{8}$ in)
American Philosophical Society, Philadelphia.

Head and shoulders, with patterned coat. On the back, a lock of white hair, lying on some platted brown hair, is framed with a surround of pearls.

This miniature was donated by the Royal Society, London, in 1943, to celebrate the 200th anniversary of the foundation of the American Philosophical Society by Benjamin Franklin.

XIV-9 Unknown artist, watercolour, after **XIV-2**, in locket.
10 × 7.6 cm (4 × 3 in)
Mary Frick Jacobs Collection, Baltimore Museum of Art, Maryland.

XIV-10 In old age by unknown artist after **XIV-2**, oil on canvas.
91.5 × 68.6 cm (36 × 27 in)

This half-length portrait shows Newton sitting in a high-backed chair with his right hand resting on a globe. Mentioned in a collection of 1742; W. Heffer and Sons, Cambridge, by 1965; sold by Thomas Agnew & Sons, London, in 1980.

XV. John Vanderbank (1694?–1739)

XV-1 Aged eighty-four, oil on canvas, 1727.
127 × 101.6 cm (50 × 40 in)
Royal Society.

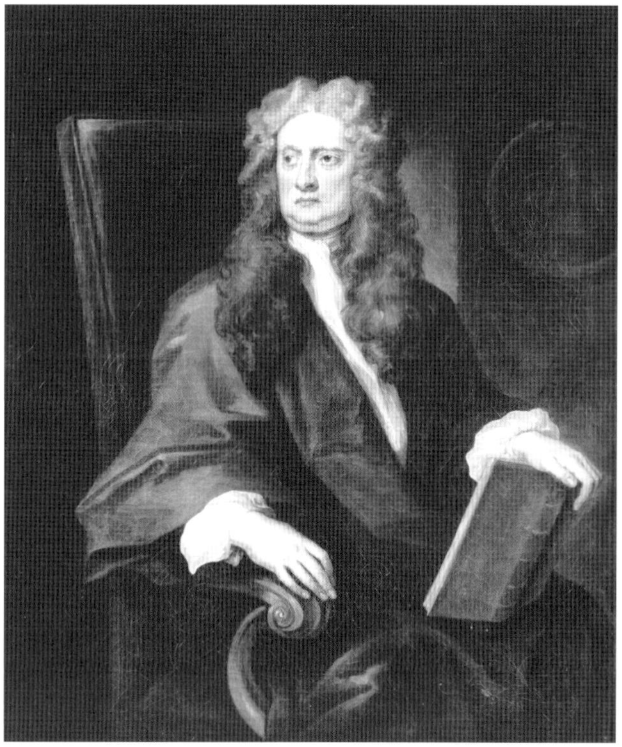

XV-1
see colour Plate VIII

Three-quarter-length life-size seated figure; body half-right, head half to his right, with long, full wig; right arm and hand resting on arm of chair, left hand holding open book on thigh; wears loose gown, soft white cravat tied under chin with ends reaching to waist inside the gown, and white shirt cuffs.; seated in high-backed chair with carved wooden arm. In the right background an *ouroboros*, a serpent with its tail in its mouth (see **XVII-1**), is painted on the wall.

On canvas, right upper background:

EQUES.
ISAACUS NEWTON.
ANNO AETATIS. LXXXV
A.C. M.DCC.XXVI

and to right of left elbow:

J. Vanderbank. Fecit
1726

After the summer break in 1726, at a time when his health was declining and he hardly went out, Newton attended only four meetings of the Royal Society and only one meeting of the Council. The last time he was well enough to make the journey from his house in Kensington into London was for the Council Meeting on 2 March 1727. When he returned home on 4 March, he was most unwell with a violent cough, and died sixteen days later. The inscriptions on **XV-1**, in which Newton looks unwell, and on the engraving (**L-1**) by John Faber, the younger (1684–1756) made from it, indicate that **XV-1** was painted when he was in his eighty-fifth year, that is, between 25 December 1726 and his death on 20 March 1727.

This portrait was presented to the Royal Society, but it is not known when, by Martin Folkes (1690–1754), PRS 1741–53, an antiquarian who had succeeded Sir Hans Sloane (1660–1753), the successor to Newton as President. In 1736 Folkes had his own portrait done by Vanderbank (who had earlier painted Newton in **XIII-2** as well as **XV-1**). In the right background of the Folkes portrait, which is in a private collection in Ireland, is a larger than life-size bust, which does not appear to show Newton's features, or to follow Guelfi's (**A.9**), or Rysbrack's (**A.10**), or Roubiliac's (**A.19**) bust of him, whatever the intent.

John Faber made an engraving of Vanderbank's portrait of Folkes in 1737 (three copies in the National Portrait Gallery) – see **Figure 4**. The bust in this portrait has long been taken to be that of Newton, that is, reflecting glory on Folkes from his association with Newton. In the catalogue at the Wellcome Library of the sale of 'mathematical instruments, gems, pictures, bronzes, busts, urns … from the collection of Martin Folkes Esq. … by Mr. Langford … on 7th and 8th May 1755', under 'Plaister Figures' (7 May, p. 3) there is 'Lot 4: A large bust of the late Earl of Pembroke [9th Earl, FRS (1693–1751)] on a painted deal term' and 'Lot 5: Ditto of Sir Isaac Newton, on a ditto' – and see **VII-1** and **A.26**. Malcolm Baker believes that Folkes's bust of

Newton was a plaster version of Roubiliac's bust **A.19**. The marble bust of the 9th Earl of Pembroke at Wilton is by Roubiliac, where there was also a plaster bust (**A.26**) by him of Newton, now missing.

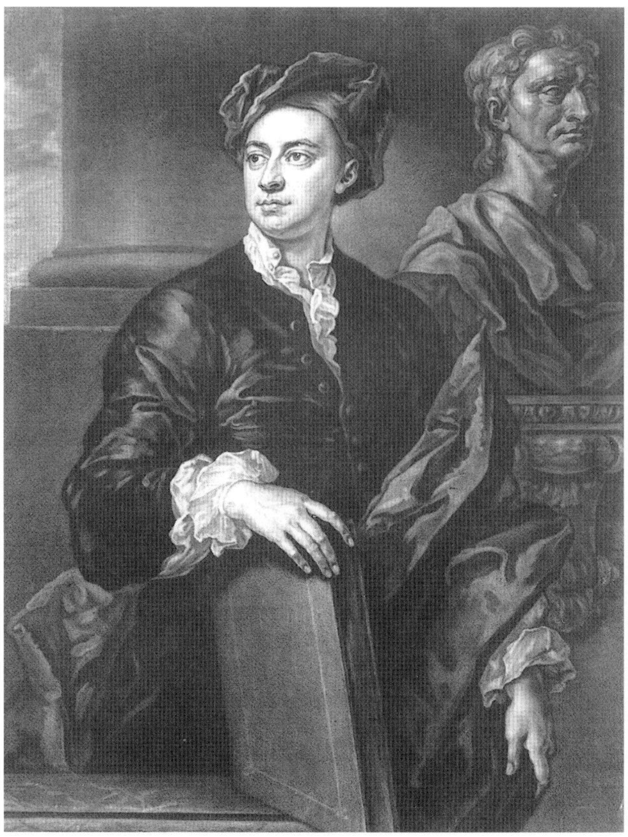

Figure 4. Engraving by J. Faber, 1737, after portrait of Martin Folkes painted by John Vanderbank in 1736, with bust in right background.

XVI. After Charles Jervas (1675?–1739)

XVI-2 Portrait of Dr Samuel Clarke (1675–1729) with bust-profile of Newton in right upper corner, oil on canvas.
127.6 × 102.9 cm ($50\frac{1}{4} \times 40\frac{1}{2}$ in)
Royal Collection, Hampton Court Palace (Millar,[23] no. 501).

Three-quarter length, wigged and facing half-right, wearing clerical costume with bands; seated, with right hand on arm of chair and left hand fingering open book; on right, on shelf behind, four volumes entitled, from left, 'BACON / OPERA', 'BOYLES / LECTURE', 'NEWTONI / OPTICA', 'NEWTONI / PRINCIPIA'; on wall above the books, in profile facing left, is sketch of head and neck of Newton, unwigged and wearing loose open shirt, with 'Sir Isaac Newton' written at base of neck; long inscription, composed by Bishop Benjamin Hoadly (1676–1761), who wrote *Essay on Life and Writings of Dr. Samuel Clarke* in 1732, and noting Clarke's death in 1729, on left above back of chair.

Millar[23] gives a date of *c.*1725 to **XVI-1**, now missing, which was painted for the lawyer Lord Lechmere (1675–1727), though it belonged to his widow when, *c.*1728, **XVI-2** was copied from it for Queen Caroline (1683–1737). Caroline favoured the low church party, and Dr Clarke, who had earlier preached a moderate form of Arianism, a doctrine for which Newton was a firm believer, was prominent at her Court. In 1716, Newton was summoned by Caroline, still Princess of Wales, to explain to her his new principles of chronology that ultimately became his *Chronology of Ancient Kingdoms Amended* published in 1728.[46]

In his will Robert Boyle, FRS (1627–1691) founded the 'Boyle lectures' to defend the Christian religion against infidels – Atheists, Deists, Pagans, Jews and Mahometans – and Richard Bentley, FRS (1662–1742), who took 'A Confutation of Atheism' as his subject, delivered the first eight in 1692. Each sermon was published separately in 1692, except for the last two

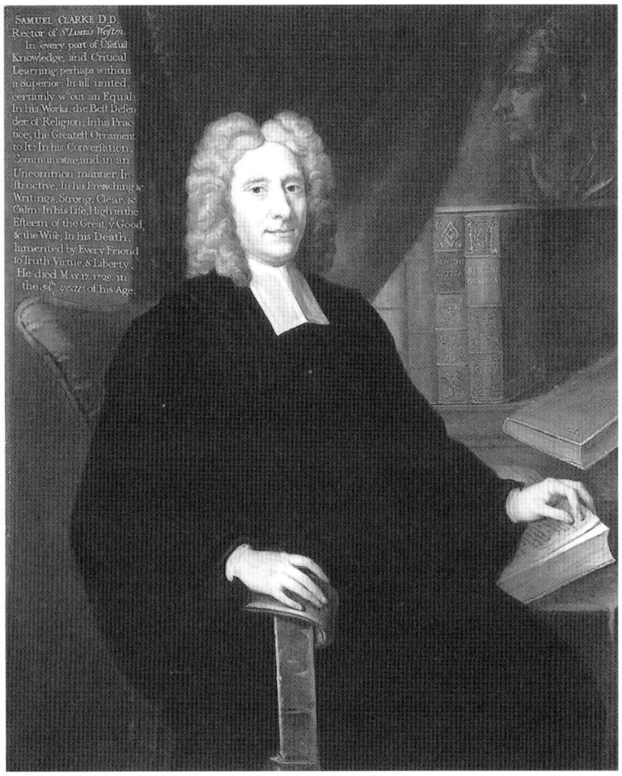

XVI-2

which were delayed to 1693 until after they had been discussed with Newton, whose 'great discovery' was used to prove the existence of an intelligent and omnipotent Creator.

Clarke gave eight Boyle lectures 'On the Being and Attributes of God' in 1704, and another course of eight lectures 'On the Unchangeable Obligations of Natural Religion, and the Truth and Certainty of the Christian Revelation' in 1705. These were at first published in two separate volumes in 1705 and 1706, but could appear bound together in corrected editions from 1706, or published as one volume of sixteen sermons (as in 1711), as indicated by the book entitled 'Boyles Lecture' in **XVI-2**. Clarke, described by William Whiston (1667–1752), a fellow Arian, as a 'bosom friend' of Newton, translated Newton's *Opticks* (1704) into Latin in 1706. Jervas had earlier painted a portrait (**VI-1**) of Newton in 1717.

Caroline set up Clarke's portrait in Kensington Palace, perhaps with the portraits of Newton after Kneller, 1689 (**II-3**, Millar[23] no. 363); John Locke, FRS (1632–1704) after Kneller, 1704 (Law no. 824; Millar no. 377); William Wollaston (1660–1724), artist unknown (Millar no. 642); and Robert Boyle by or after Johann Kerseboom (d.1708), 1689, that had belonged to George I (Law[23] no. 843; Millar no. 332). Later, in her Grotto at Richmond she placed marble busts of Clarke, Newton, Locke and Wollaston centred on one of Boyle, that can all be safely attributed to Giovanni Baptista Guelfi – see **A.7**, **A.8** and **A.9**.

XVII. William Hoare, RA (1707?–1792)

XVII-1 Profile drawing of Newton in red chalk, 1734.
30.2 × 26 cm ($11\frac{7}{8} \times 10\frac{1}{8}$ in)
British Museum since 1894 (1894-12-19-1).

Head, in profile to left, with short curling hair, a flaming comet in background, treated as a sculptured medallion surrounded by the eternal circle of a serpent with its tail in its mouth, the *ouroboros*, the alchemical tail-devouring symbol implying eternity because of constant regeneration, and with eternity signifying wisdom. The *ouroboros* also correlates Newton's alchemy with his interest in the revelations of antiquity. It featured in the Vanderbank portrait **XV-1** of 1727.

Hoare made an etching (**LVII-1**) of this drawing the same year. Oliver Goldsmith (1730?–1774) in his 1762 biography[15] of Beau Nash (1674–1761) of Bath, wrote:

> To add to his [Nash's] honours, there was placed a full length picture [by William Hoare] of him, in *Wiltshire's ball-room*, between the busts of *Newton* and *Pope*. It was upon this occasion that the Earl of *Chesterfield* [Goldsmith's attribution] wrote the following *severe*, but *witty* epigram:
>
> Immortal *Newton*, never spoke
> More truth, than here you'll find:
> Nor *Pope* himself, e'er penn'd a joke
> More cruel on mankind.

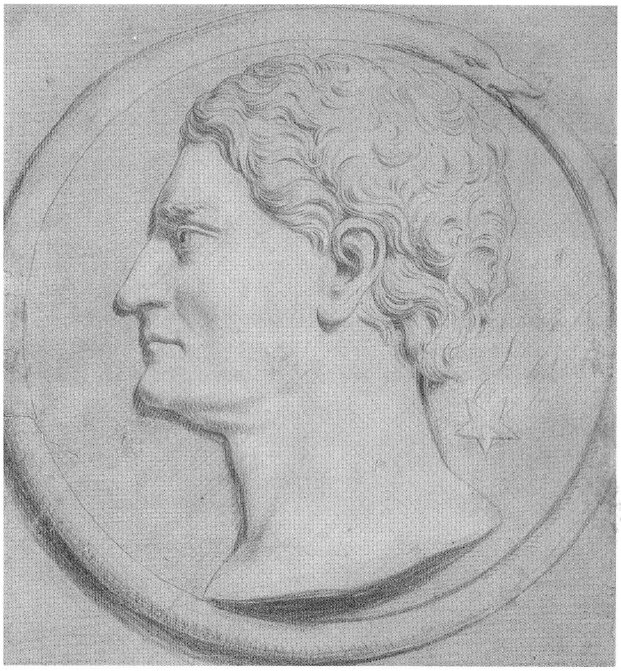

XVII-1
see colour Plate XI

> This picture plac'd the busts between,
> Gives satyr all his strength;
> *Wisdom* and *Wit* are little seen,
> But *Folly* at full length.

The epigram was quoted from The *Gentleman's Magazine* 1741; 11: 102, and see Wimsatt.[39] It would be of interest to be able to identify the bust of Newton mentioned: it is tempting to suggest that it and the bust of Pope were the same two busts carved by Roubiliac owned by Earl Poulett at Hinton St George about 40 miles from Bath, and sold at Sothby's in 1968 – see **A.22**.

XVIII. Louis-Fabricius Dubourg (1693–1775)

XVIII-1 Drawing for engraved frontispiece of book by F M Arouet de Voltaire (1694-1778), *Elémens de la Philosophie DE NEUTON* par MR. DE VOLTAIRE, 1738[34] – see engraving **LVIII-1** below.

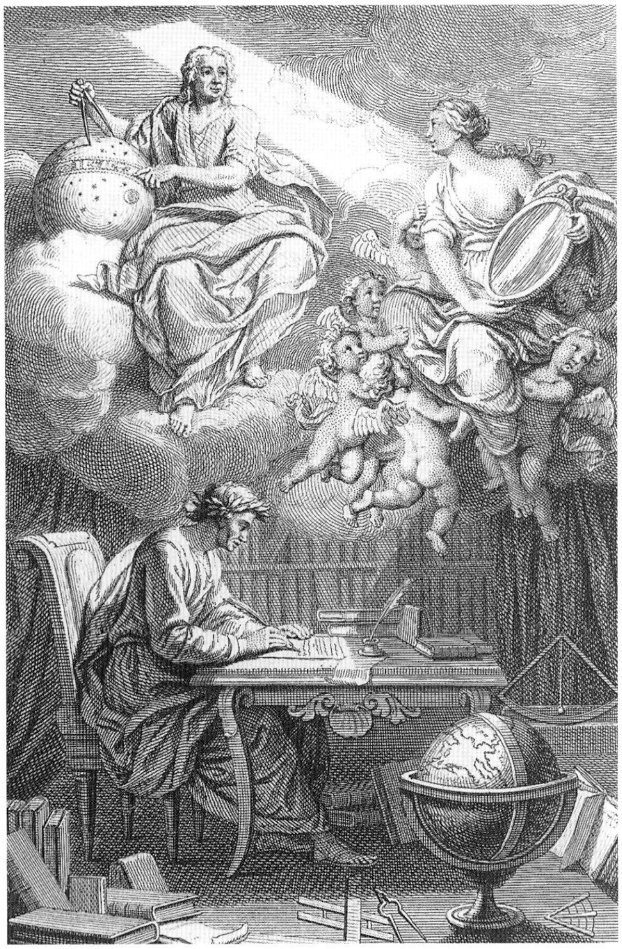

LVIII-1

The identification of the three figures in this allegorical drawing, typical of the period, is uncertain. Above, on the left, on a throne of clouds sits a man in classical garb, in his right hand a compass resting on a celestial globe. With the index finger of his left hand he points to a band encircling the globe with the signs of the zodiac. The figure could represent the Muse Urania, the muse of astronomy, with her usual attributes of globe and compasses, but bears a strong resemblance to Newton, perhaps suggesting he has the status of a muse for others. To the right, is a woman with a bared left breast, held aloft by putti and holding a mirror, which reflects a shaft of divine light (or light of reason) on to a seated figure below. Below, the seated figure, robed in a Roman toga, sandals on his feet, has a laurel wreath on his head, and sits at a table, pen in hand. Behind, are shelves of books, and in front on the floor is a terrestrial globe, with more books and scientific instruments. The writer, despite the garland, globe and instruments, seems to be Voltaire who is intended. The bare-breasted woman has the appearance of truth, or philosophy, or science because of the mirror, but as well suggests education. Walters[34] has argued that she is also Madame du Châtelet (1706-1749), mathematician and mistress of Voltaire, who began translating the *Principia Mathematica*[25] from Latin into French in 1745, the translation being posthumously published in 1759.

In mentioning Madame du Châtelet, at the top of the frontispiece to her book *Institutions de physique*, Paris, 1740, are three oval portraits that have been claimed to be of Descartes, Newton and Copernicus, while the main part below has been said to depict Newton mounting the temple of truth. Both images are too uncertain to be accepted as likenesses of Newton. Likewise, an eighteenth-century allegorical fresco in the Klementinum in Prague, in which a female muse holding a telescope is claimed to be Newton, cannot be accepted.

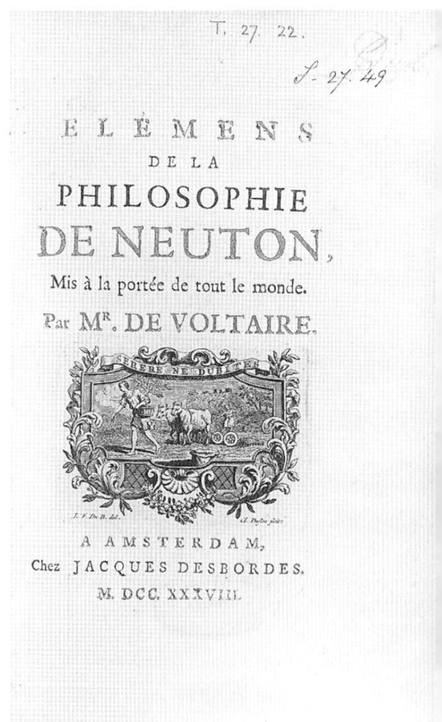
ELÉMENS
DE LA
PHILOSOPHIE
DE NEUTON,
Mis à la portée de tout le monde.
Par Mr. DE VOLTAIRE.

A AMSTERDAM,
Chez JACQUES DESBORDES.
M. DCC. XXXVIII.

XIX. Thomas Hudson (1701–1779)

XIX-1 Posthumous portrait, oil on canvas, date unknown.
210.8 × 132 cm ($83\frac{1}{4} \times 52$ in)
Trinity College, Cambridge.

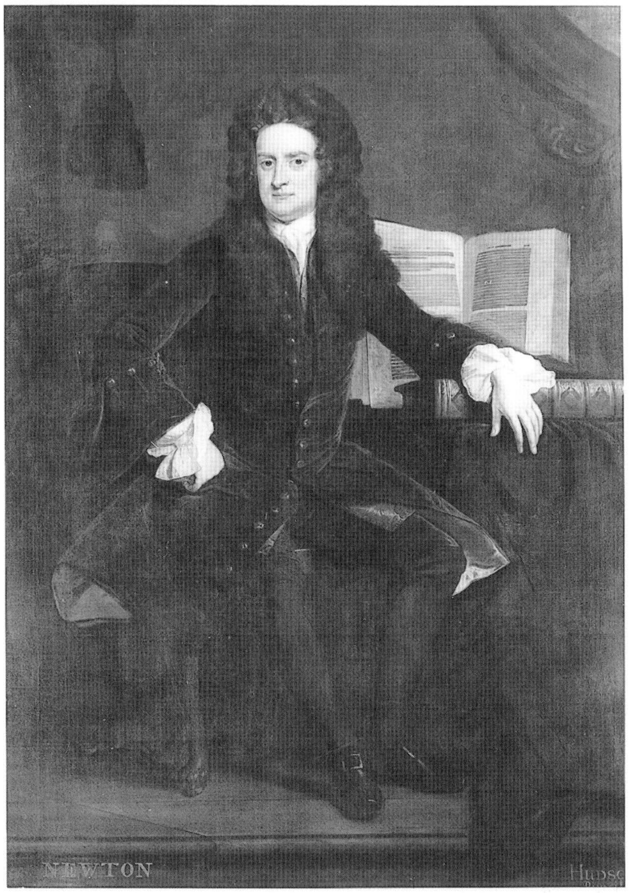

XIX-1

Full-length, life-size seated figure; head slightly left, eyes to front, clean shaven, wearing a long dark wig; body slightly right; right hand rests on thigh, left hand on large closed book on table; wearing brown coat, white cravat and cuffs, brown knee breeches and stockings, black shoes with gilt buckles; seated in high-backed chair; table on right with two folio volumes, one closed, the other open.

Low left on canvas:

NEWTON

and low right:

HUDSON pinxit

Hudson was twenty-four when Newton died at eighty-four in 1727. The head appears to be taken from the Kneller portrait of 1702 (**III-1**) when Newton was fifty-nine.

XX. Francis Hayman (1708–1776)

XX-1 Painting of Newton in cove of ceiling above main staircase and landing, 1743.
Little Haugh Hall, Norton, Bury St Edmunds, Suffolk.

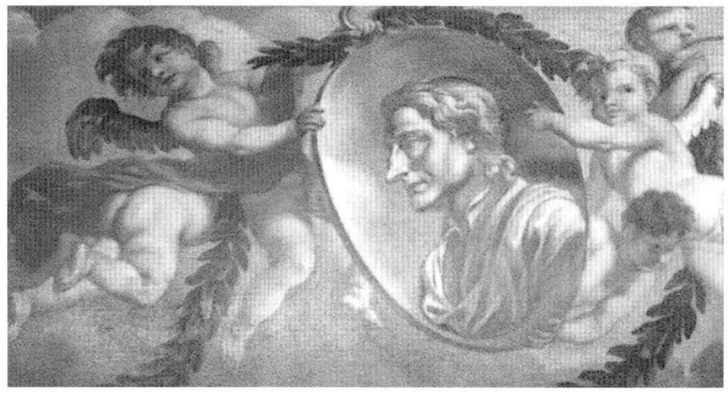

XX-1

The central oval of the ceiling, depicts *Apollo and the Muses crowning Archimedes with a laurel wreath.* In the 12 inch high, curved junction, or cove, between the wall and the ceiling, there are bust-portraits of Galileo (?), Newton and Nicholas Saunderson, FRS (1682–1739), who was blinded in infancy from smallpox, mathematician made a member of Christ's College, Cambridge, in 1707 and Lucasian Professor of Mathematics in 1711. These portraits are surrounded by putti, with garlands and instruments, or tokens, of astronomy, physics and mathematics after the relief on Newton's monument in Westminster Abbey by Rysbrack (see **D.3**). Newton's head is painted in profile, with shoulders loosely draped, in an oval plaque supported by the putti. Little Haugh Hall was modernised and decorated for Dr Cox Macro (1683–1767), antiquary, chaplain to George II and friend of Rysbrack, who was at Christ's College as an undergraduate, but there too late to have known Newton.[1]

XXI. Arthur Devis (1712–1787)

XXI-1 *The John Bacon Family*, oil on canvas, *c.*1742–3.
76.5 × 131.2 cm ($30\frac{1}{8} \times 51\frac{5}{8}$ in)
Paul Mellon Collection, Yale Center for British Art (57-61.27), New Haven.

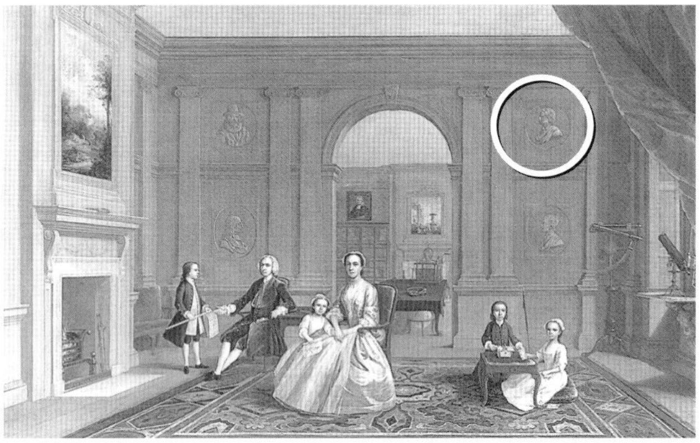

XXI-1

John Bacon, FRS (1709–1752) and his wife Catherine are with their four children, sitting with their backs to a wall, on which there are four oval, grisaille medallion bust-length portraits of Sir Francis Bacon and Sir Isaac Newton, above, and John Milton and Alexander Pope, below, each set within a painted rectangle in grey monochrome. Through an arch in the wall, scientific instruments can be seen resting on a table, which has twinned terrestrial and celestial globes under it, and with a portrait of Edmond Halley by Thomas Murray, *c.*1712, in the background. To the right, by a window is a transit quadrant for determining solar and sidereal time and a reflecting telescope.[39]

Provenance: By descent from John Bacon; Sir Percy Loraine; with Spink & Son as an *Unknown Family*, from whom acquired in 1968 by Paul Mellon.

John Bacon was not a relative of Francis Bacon. The profile bust portrait of Newton in this oil appears to be after Rysbrack's Conduitt bust **A.10**.

XXII. Unknown artist

XXII-1 Emblematic portrait, in grisaille with a dark terracotta background, oil on canvas, *c.*1750.
75.5 × 137 cm ($29\frac{3}{4} \times 54$ in)
Government Art Collection (GAC 9267) since purchased from F T Sabin in 1971.

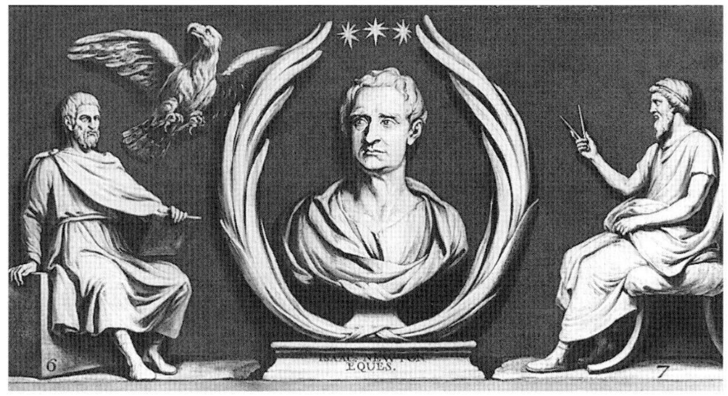

XXII-1

A bust of Newton, after Roubiliac **A.19**, is in the centre, surrounded by a garland wreath and inscribed on its base:

ISAAC NEWTON
EQUES.

Above Newton's head are three stars. On the left of the garland is a seated figure with a divining rod in his left hand and with a soaring eagle to the right of his head. To the right of the garland is another seated figure, a mathematician with a pair of compasses in his right hand.

Exhibited at an exhibition of English Portraits 1500–1830 at Sabin Galleries, London, in 1970. Probably once part of a library decoration.

XXIII. Biagio Rebecca (1735–1806)

XXIII-1 *Trompe-l'œil* medallion portrait relief *en camaïeu*, variant of the Delvaux portrait medallion (**B.16–B.22**), oil on canvas, *c.*1772 (?).
Oval, 71 × 60.5 cm (28 × 23¾ in)
In Entrance Hall, designed by Robert Adam (1728–1792) in 1767, of Osterley Park, Middlesex (National Trust).

Inscribed:

ISAAC NEWTON ANG. NAT. 1642. OB. 1727.

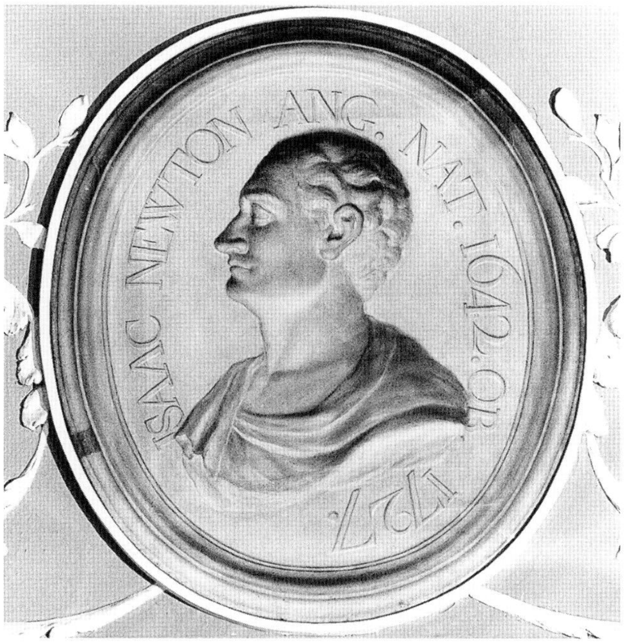

XXIII-1

Such *trompe-l'œil* monochromes were a speciality of Rebecca, who did some for Robert Adam at Kedleston Hall in the 1780s. 'ANG' is an abbreviation for *Anglus* or Englishman, suggesting that the placement of this *trompe-l'œil* was patriotic, and, with the Triumph of Ceres and the Triumph of Bacchus, the two other ones in the Hall (which also appear to be by Rebecca, though they have been attributed to G. B. Cipriani in the past), emphasises the benefits conferred on mankind by these three. These monochromes (unlike Zucchi's insert paintings in the Library) are not listed in the Osterley inventory of 1782, so they may be later additions to the original scheme of 1767–72.

XXIV. Giovanni Battista Cipriani (1727–1785)

In the will of Dr Robert Smith, FRS (1689–1768), Master of Trinity College, Cambridge, from 1742 to his death, the sum of £2,000 was bequeathed to the 'master and Seniors' for 'such publick Uses or Ornaments in the College as they think proper; only desiring that the Window in the South End of the College Library be glazed with the best painted glass'. The great window that dominates the south end of the Wren Library was designed by Giovanni Battista Cipriani and completed by William Peckitt in 1775. It is one of the few additions made to Wren's new library, begun in 1677 and completed in 1692, into which the books from the old library in Great Court were moved in 1695 (Junior Bursar's accounts, 1694/5).[22,30]

XXIV-1 Design for the south window in the Wren Library, Trinity College, in pen and black ink, watercolour, bodycolour and traces of graphite on paper, 1771–2.
50.8 × 23.8 cm (20 × 9⅜ in) with arched top
Sketch presented to the Fitzwilliam Museum by Charles Fairfax Murray in 1912 (no. 738/25).

This bright and unfaded sketch shows Newton being presented to a man, who sits under a canopy attended by the British Minerva, or Britannia, wearing a laurel wreath and holding the Laurel Chaplet in his outstretched hand. In the right foreground, Sir Francis Bacon, wearing a hat, sits with an open book on his lap and a pen in his right hand. He is seated 'as if preparing to register the reward about to be bestowed on Sir Isaac' (*A Guide through the University of Cambridge*, Cambridge, 1803, p. 96). The bare-breasted woman in yellow presenting Newton, as in **XXIV-4**, has been variously described as Fame, or the Muse of the College. The window was commissioned in 1771, and this is the preliminary sketch for it (see p. 44).

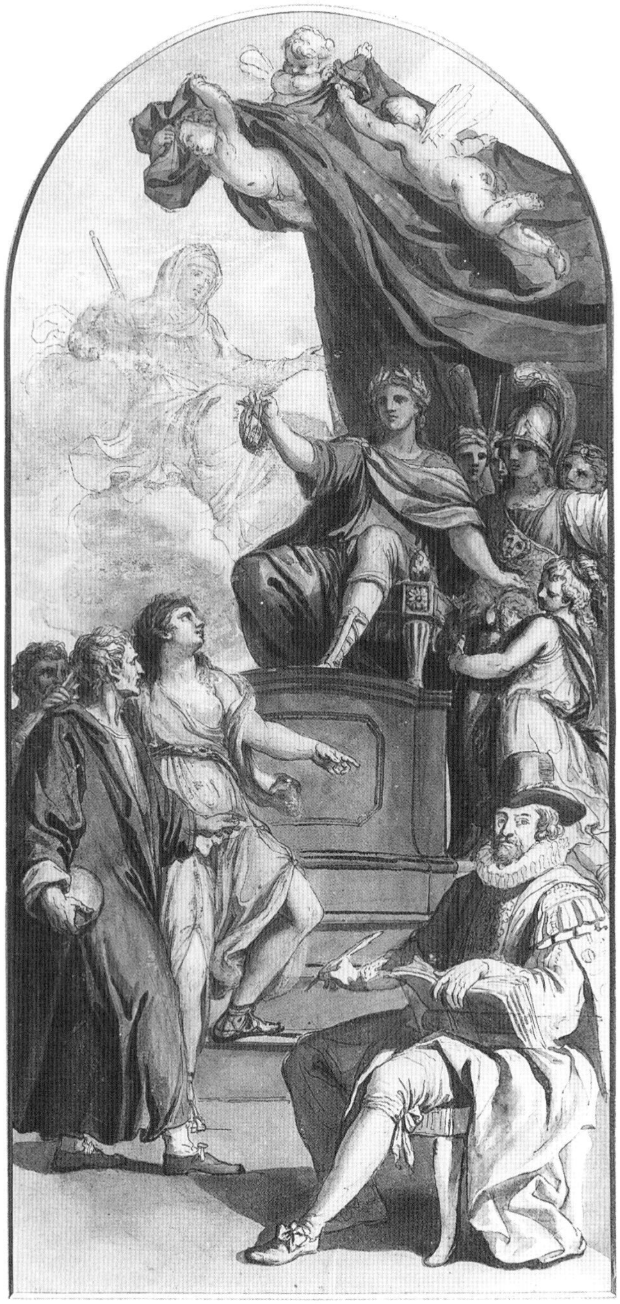

XXIV-1 Preliminary sketch for the south window in the Wren Library, Trinity College, Cambridge

XXIV-2 Watercolour for the south window, Wren Library, 1772–3.
65 × 34 cm ($25\frac{1}{2} \times 13\frac{1}{2}$ in) with arched top
Wren Library, Trinity College, Cambridge.

In this watercolour there is a change from the sketch, and it is the contemporary George III who sits on a throne under the canopy, holding the Laurel Chaplet in his right hand. Newton is pointing with his left hand, explaining the sphere held in his right hand. Francis Bacon, now in his full Chancellor's robes and no longer wearing a hat, 'is, by an anachronism legitimate in art, proposing to register the reward about to be conferred upon Sir Isaac'.[6] *The New Cambridge Guide*, 2nd edn, Cambridge, 1812, p. 71, explains the anachronism: 'It must be allowed the incongruity of the design is as absurd as could well be imagined; but the artist being a foreigner, may in some degree stand excused for the ignorance of English history, displayed by the introduction of three characters, who never were contemporaries, into one picture.'

There is an alteration to the watercolour made by the pasting on to it of a piece of paper on which the King's face and the crown have been repainted. The watercolour has not been squared up as it would have been if used for transference to the full-sized cartoon (**XXIV-3**). This suggests that it provided a record of the model used to make the cartoon, and is likely to have been done before the glass window was made, a point reinforced by the colouring of this faded watercolour, which is drab compared with the bright colours used for the glass.

A payment of £105.0s.0d was made to Cipriani in 1773 (Senior Bursar's accounts, 1772/3), implying that he finished the commission that year, including the cartoon, although the alteration to the crown in the cartoon (and this watercolour) may have been made in 1774, or even 1775. On the back of the watercolour is stuck a Robert Smith legacy bookplate of 1768, on which is written: 'John Baptista Cipriani Deleate it [*sic*], 1775', and that the costs for the window were met from money bequeathed to the College by Smith. Despite the differences in **XXIV-2** from the cartoon (**XXIV-3**) and the window (**XXIV-4**), and despite the similar repainting of the crown on it and the cartoon, the date '1775' on the bookplate has in the past led to the conclusion that the watercolour followed after the window was finished in 1775, which has here been argued against.

XXIV-3 Full-size cartoon for the south window in the Wren Library, in red and black chalk, charcoal under-drawing strengthened with black watercolour/brown ink, on many joined sheets of paper; squared up with red chalk; lines relating to architectural elements added using graphite and white chalk, 1772–3.
5.03 × 2.51 m ($16\frac{1}{2} \times 8\frac{1}{4}$ feet)
Wren Library, Trinity College, Cambridge.

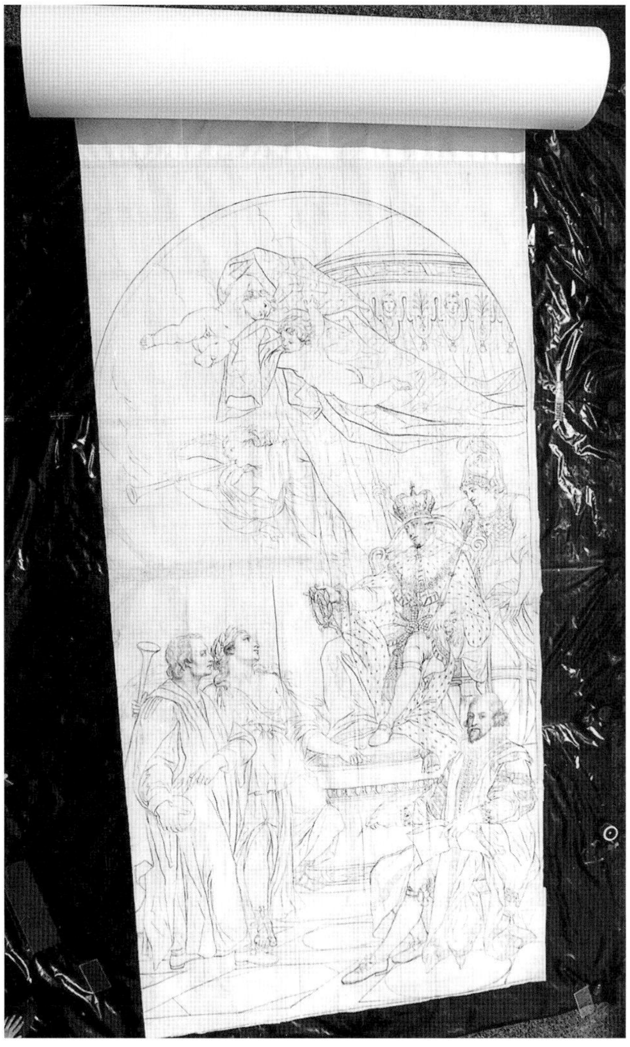

XXIV-3

The full-size cartoon for the window follows the watercolour (**XXIV-2**) except in the drawing of Bacon's left foot, the decorations hanging down from the domed top of the tent to which the canopy is attached, and the position of the Muse's left arm and hand; the King's crown, but not his face, has also been repainted in the same way as done to the watercolour with the pasting on to it of a patch of paper. The cartoon was repaired with full conservation in 1997.

XXIV-4 Glass window at the south end of the Wren Library, coloured using a mixture of enamel and pot metal with yellow stain, completed in 1775 by **William Peckitt of York (1731–1793)**.
5.03 × 2.44 m ($16\frac{1}{2} \times 8$ feet) – there are $16\frac{1}{2}$ panes of glass to the full height and 8 panes across the width of the window, an area of about 136 square feet.
Wren Library, Trinity College, Cambridge.

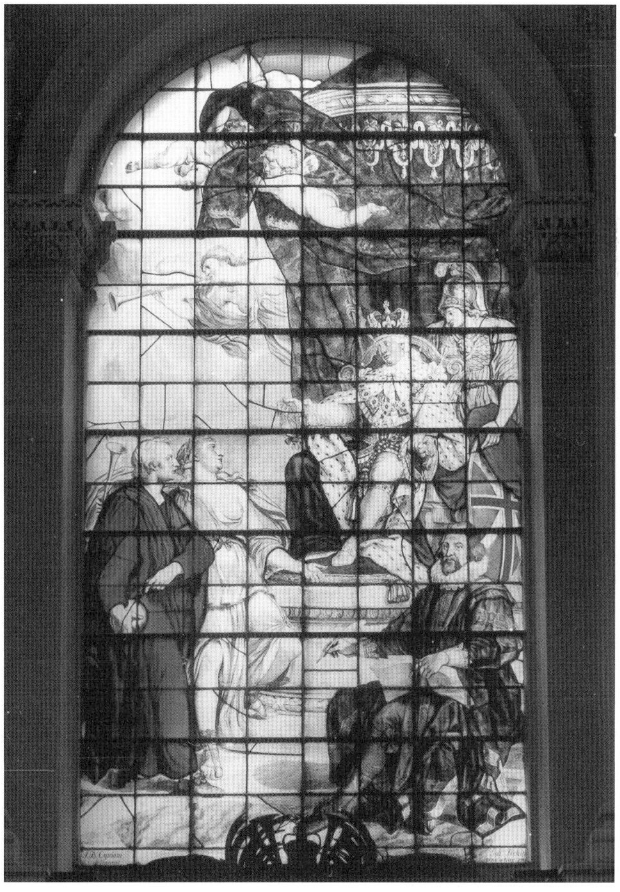

XXIV-4
see colour Plate XI

The brightly coloured glass window depicts the presentation of Sir Isaac Newton to George III, in the presence of Sir Francis Bacon. Behind the King is the British Minerva, or Britannia, holding a shield of the Union of England and Scotland. As in **XXIV-1**, the bare-breasted female, holding a trumpet, clad in deep yellow, and presenting Newton to the King, has been variously described as Fame and (more convincingly) as the Muse of the College. The King in his parliamentary robes sits crowned on a raised, canopied Chair of State, holding a sceptre in his left hand and proffering the Laurel Chaplet to Newton with his right hand. Two cherubs draw back the curtains of his canopy, and a winged angel blows her trumpet in the background.

Below, at bottom left, is written:

J. B. Cipriani
delineavit

Below, at bottom right, is written:

Guli^{us} *Peckitt*
pinx & tinx 1775

On completion of the window in 1775, William Peckitt was paid the sum of £315.0s.0d 'for painting and staining the south window of the Library' (Senior Bursar's accounts, 1774–5; William Peckitt's Account Book in York City Art Gallery; William Peckitt's Commission Book[7]). The window was restored in 1970.

XXV. Januarius Zick (1730–1797)

Allegorie auf Newtons Verdienste um die Gravitationslehre. ('Allegory of Newton's contribution to the theory of gravity'), oil on canvas, *c.*1780.
63 × 73 cm ($24\frac{3}{4}$ × $28\frac{3}{4}$ in)
Niedersächsische Landesgalerie, Hanover.

The picture is an appraisal of the age of enlightenment and modern science represented by Newton, who is postured between the offerings of a bacchanalian life and of a science that is firmly based on the ground. He is shown standing awkwardly at the base of a monument, topped by an image of Pegasus, on the slopes of Mount Parnassus, his left arm stretched towards a proffered goblet and with his right hand gesturing at a terrestrial globe and pile of books placed on the ground in front of him. To his right, the nine muses play instruments, and in the right background are ruined buildings of classical antiquity.

There is a second painting with the same measurements, also by Zick, in this gallery in Hanover, entitled *Allegorie auf Newtons Verdienst um die Optik* ('Allegory of Newton's achievements in optics'). This has a man, said to be Newton, standing with one foot resting on the personification of falsehood with its mask, hair made of snakes and torch, depicted as a fat man lying on his side, and pointing to two men, said to be Euclid and Diogenes, stepping away from a temple, in which Amor stands in front of the altar. Above, a flash of lightning comes out of the clouds, and in front of the two men is a sundial showing the first hour. This painting is not accepted as an allegory involving Newton: Francis Haskell in 'The apotheosis of Newton in art' in *The Texas Quarterly* 1967; 10: 218-37 did not accept that either painting by Zick referred to Newton's discoveries.

XXVI. Ascribed to Jacob de Wit (1695–1754)

Head and shoulders in profile, after L. Delvaux (**B.16**), oil on panel painted *en camaïeu* (painted imitation of a cameo).
Oval, 42 × 33 cm ($16\frac{1}{2}$ × 13 in)
Private collection.

Inscribed
ISAAC NEWTON ANGLUS NATUS 1642
OBIIT 1727
round edge.

First recorded in posthumous inventory of the collection of Markgräfin Karoline Luise in 1784 without attribution, but ascribed to Jacob de Wit in Jan Lauts, *Katalog alte Meister bis 1800*. 2 vols. Karlsruhe: Staatliche Kunsthalle, 1966, vol. I, no. 242, pp. 313–14, vol. II, plate on p. 434. This painting is after the relief medallion of Newton (**B.16**) by **Laurent Delvaux (1696–1778)**.

XXVII. Attributed to Biagio Rebecca (1735–1806)

XXVII-1 Portrait in ceiling roundel, after Kneller portrait of 1702 (**III-1**), painted *en camaïeu* between 1781 and 1784.
30.5 cm (12 in) in diameter
The Library, Berrington Hall, Herefordshire (National Trust).

Head, in full wig with open shirt, white on blue ground, within painted frame.

Biagio Rebecca is reputed to have decorated the ceiling of the drawing room, as well as the ceiling and paintings on the walls of the library at Berrington Hall. There are eight ceiling roundels of Matthew Prior, John Milton, Alexander Pope, William Shakespeare, Geoffrey Chaucer, Isaac Newton, Francis Bacon and Joseph Addison in that order. Painted with emphatic shadows, these medallions were meant to look as much like stucco in high relief as possible. The house was built by Thomas Harley (1730–1804), son of the 3rd Earl of Oxford, and, after descent in the Rodney and Cawley families, was transferred to the National Trust in 1957.

XXVIII. Pieter de Gree (1751–1789)

XXVIII-1 Portrait of Newton 'in Antique style', after the Delvaux bust relief **B.17**, formerly in Musées Royaux des Beaux-Arts de Belgique in Brussels, now missing.

Reynolds praised de Gree, who worked particularly in Ireland, in 1785 as 'a very excellent painter in chiaroscuro in imitation of basso-relievos. He paints likewise portraits in oil and in crayons extremely well.' Present whereabouts of the portrait is unknown, but is it the oil grisaille painting, of which there is a record and photograph in the Archives of the National Portrait Gallery when it belonged to J. O. Feather of London in 1974? The inscription on it, however, differs from that of the Delvaux, to which it is otherwise very close, in having 'ISAAC NEWTON ANGLUS AETAT SUAE42 OBIIT 1727' and not 'ISAAC NEWTON ANGLUS NATUS 1642 OBIIT 1727'.

XXIX. Giovanni Martino de Boni (1752 – after 1810)

XXIX-1 Allegorical painting *Britannia between Newton and Willem Jacob 'sGravesande*, oil on canvas, 1787.
80 × 65 cm ($31\frac{1}{2} \times 25\frac{1}{2}$ in)
Galleria Doria Pamphilj, Rome (FC 648)

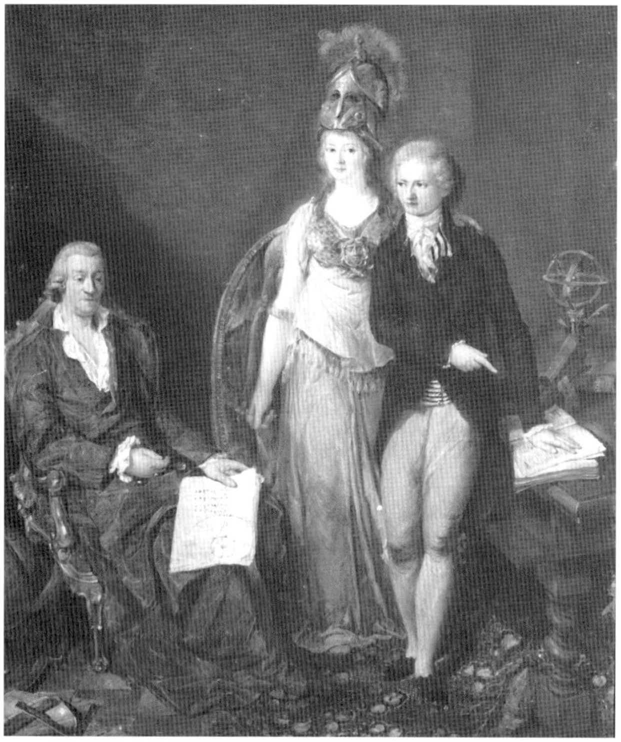

XXIX-1
see colour Plate XII

An elderly Newton sits in a chair on the left of the painting, holding in his left hand a piece of paper on which there is a diagram and several equations. Britannia, fully helmeted, stands with her shield in the centre, and has her left hand on the shoulder of a smartly dressed young man. His left hand is placed flat on an open book lying on a table on which a mathematical instrument and other books are piled. Partially under the open book on which the young man has his hand is a scroll of paper on which is written *Sig. Gravesand.*

Signed and dated:

Gio. Mar. de Boni 1787

No. 107 in *The Romantic Movement*, fifth exhibition to celebrate the tenth anniversary of the Council of Europe, by the Arts Council of Great Britain, Tate Gallery, London, 10 July – 27 September 1959. The Dutchman Willem Jacob 'sGravesande, FRS (1688-1742) was the earliest influential exponent of the Newtonian philosophy in continental Europe. His book *Physices elementa mathematica, experimentis confirmata. Sive, introductio ad philosophiam Newtonianam* was first published in Leiden in 1720.

XXX. George Romney (1734–1802)

XXX-1 'Newton displaying the prism', six drawings in graphite, *c.*1792.
49.2 × 38.6 cm ($19\frac{3}{8} \times 15\frac{1}{4}$ in)
Department of Prints and Drawings, British Museum (1914-1-17-23-28), acquired in 1914.

Six sketches in a collection of fifty-six drawings, mostly in graphite, put together in an album, and taken from sketchbooks once in the possession of Miss Romney, sold Christie's 10 May 1834 (lot 77) and 25 May 1894 (lot 204).

XXX-2 Pages from another sketchbook with several composition trials for 'Newton displaying the prism', *c.*1792.
13.2 × 18 cm ($5\frac{1}{5} \times 7$ in).
Sold Christie's 25 May 1894 (lot 204); J. P. Heseltine; Lord Nathan of Churt. Exhibited in *The European Fame of Isaac Newton*, an exhibition at the Fitzwilliam Museum, Cambridge, 1973–4.[10]

XXX-3 'Newton displaying the prism', pen and brown ink on paper, *c.*1792.
14.8 × 18.9 cm ($5\frac{13}{16} \times 7\frac{7}{16}$ in)
Yale Center for British Art (B1979.1)

XXX-4 'Newton displaying the prism', oil on canvas, 1794. 157.5 × 221 cm (62 × 87 in)
Romney sale as 'Sir Isaac Newton contemplating the Phaenomenon of the refraction of Rays by the Prism', Christie's 27 April 1807 (lot 118); Revd John Romney; Christie's 29 June 1810 (lot 25); William Chamberlayne (d.1829); by descent in Chamberlayne family at Cranbury Park, near Winchester; Scarsdale Sale at Christie's 18 July 1930 (lot 84); in collection of Mrs Chamberlayne-MacDonald, when exhibited (no. 38) at The Iveagh Bequest, Kenwood, London in 1961.

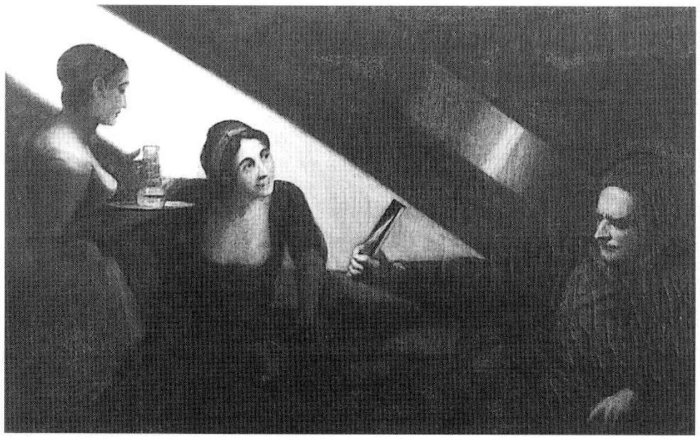

XXX-4

The description at the Scarsdale Sale in 1930 was as follows: 'Newton is seated on the left intently observing the prism, which he holds in his extended right hand in a sunbeam; two young ladies (Romney's nieces) opposite him, one astonished at the phenomenon of the prismatic colours projected on the wall, whilst the other is in the act of carefully bringing a carafe of water; the countenance of one of the ladies is in shadow, and the other (the smiling girl) in a reflected light.'

Signed: '*Romney*', and dated 1794.

The painting is based on the light-splitting experiment of 1666. According to Romney's son, Revd John Romney, it was painted from the death mask of Newton at Trinity College, Cambridge (**C.2**), the same mask that Roubiliac had used as a model for the head of his statue (**D.7**) there. There is a stipple engraving (19.5 × 27 cm) of it by R. M. Meadows (d.1812), 14 April 1809, in the Wellcome Library.

XXXI. William Blake (1757–1827)

XXXI-1 Drawing of Sir Isaac Newton, pencil, 1795. 20 × 25.5 cm ($7\frac{7}{8}$ × 10 in)
Fitzwilliam Museum, Cambridge, since 1982 (Keynes Family Trust).

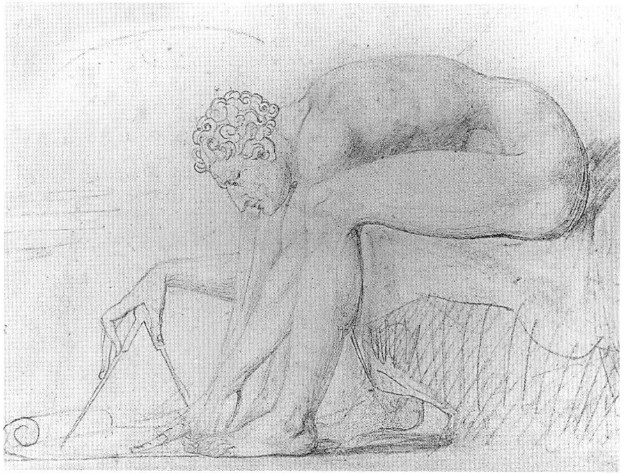

XXXI-1

The naked figure of Newton sits facing to the left and bending forwards to a scroll on the ground in front of him, with a pair of compasses in his right hand. His pose is developed from Abias, one of Michelangelo's frescos of the Ancestors of Christ on the ceiling of the Sistine Chapel, which Blake had copied from an engraving in a drawing now in the British Museum.

Blake was conversant with Newton's theories of colour, but thought that, on the grounds of his power of reasoning, Newton was the spirit of materialism. As well, he asserted that Bacon, Locke and Newton were the three great teachers of Atheism, or Satan's doctrine.[4,19] Exhibited at the Fitzwilliam Museum 22 November 1973 – 6 January 1974.[10]

The drawing was preliminary for the coloured monotype (**XXXI-2**) now in the Tate Gallery.

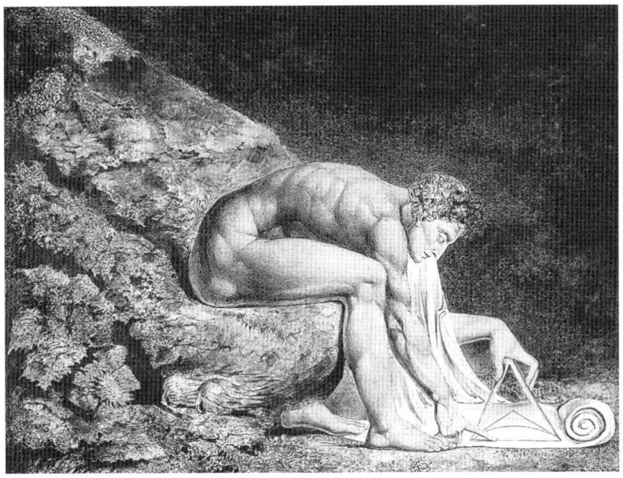

XXXI-2

XXXI-2 *Newton*, colour-printed monotype finished in pen and watercolour, 1795.
47.9 × 60.3 cm ($18\frac{7}{8} \times 23\frac{3}{4}$ in)
Tate Gallery, London.

Newton faces to the right in reverse of the drawing, and appears to be seated at the bottom of the sea, the water symbolising the materialism of Newton's philosophy, and the pair of compasses the symbol of material creation. Newton is the self-obsessed rational man. Signed '1795 W B inv [as monogram]', and inscribed 'Newton' below the design.

One of a series of twelve designs, which Blake described as being of 'Historical and Poetical' subjects, executed *c.*1795.[4] There is a second copy of this colour-print in Philadelphia with an earlier watermark than the one in the Tate.

DOUBTFUL AND SPURIOUS PORTRAITS

XXXII. Unknown artist, but ascribed to Sir Godfrey Kneller (1646–1723)

XXXII-1 An old man, oil on paper laid down on panel.
43.2 × 28 cm (17 × 11 in)
Earl of Portsmouth, Farleigh Wallop, Basingstoke, Hampshire.

Three-quarter-length; full face with large staring dark eyes; body turned to left, seated, with right hand resting on table holding a quill pen, and left hand placed over right wrist; wearing a gown with white shirt open at neck, and somewhat wild-looking long hair down to shoulder; there is a crumpled sheet of paper lying on the table in front of his right hand.

High on left is written:

SIR ISAAC NEWTON
By Kneller

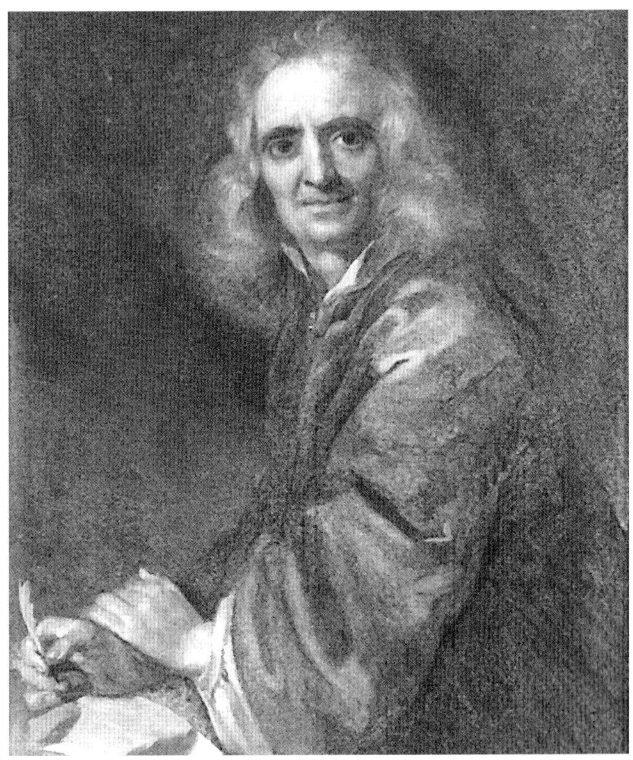

XXXII-1

By descent with Earls of Portsmouth; placed in the Viscount Lymington Sale of the Newton Papers at Sotheby's in 1936, but withdrawn before the sale.

The painting, if of Newton, only partially confirms Conduitt's description (Keynes MS 129A, pp. 23–4, in King's College Library, Cambridge) of the aged Newton, that he had 'a very lively piercing eye, a comely and gracious aspect. His head of hair, as white as snow, was full with no baldness. Even as an old man he retained the bloom and color of youth and all his teeth except one.' More obviously the remark of Bishop Atterbury (1662–1732)[2] about Newton in his last years is suggested: 'Indeed, in the whole air of his face and make, there was nothing of that penetrating sagacity which appears in his composures. He had something rather languid in his look and manner, which did not raise any great expectation in those who did not know him.' In this portrait of an old man, the hair is grey rather than white, flecked with auburn, and very thin over the dome of the head showing baldness. Newton's eyes were blue and not dark, and his hair never auburn.

The painting, cleaned in 2000, is not mentioned by J. D. Stewart.[31] From a photograph Stewart believes that the portrait is not by Kneller on grounds of style, as the character of the design and the lack of idealisation and friendliness in the features are not those found in Kneller's works. He suspects that, if the image does represent Newton, it is probably posthumous (which precludes Kneller), but in any case the unsigned portrait remains unlikely to be an image of Newton, despite its provenance.

XXXIII. William Gandy, junior (*c.*1655–1729)

XXXIII-1 Portrait, oil on canvas, 1706.
73.7 × 61 cm (29 × 24 in)
Collection: John Abel Walter of Ashley Mount, near Lymington; Margadale Collection sold Christie's 10 December 1971 (lot 84).

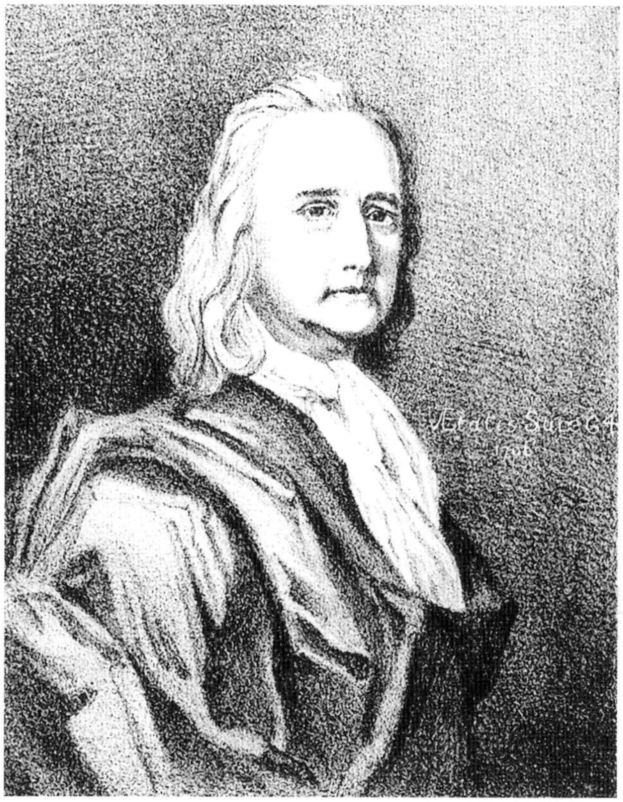

XLIV-1

Half-length, body facing to the right, with head half-turned to his right and shoulder-length hair, wearing brown coat and white cravat.

Inscribed and dated, left centre:

AEtatis Suae 64
1706

We only know of the portrait, said to be of Newton in his sixty-fourth year and painted by William Gandy in 1706, from the lithograph (**XLIV-1**) done in 1848. In this print it is not possible to discern Newton's features.

XXXIV. Unknown artist

XXXIV-1 Portrait said to be of Newton, oil on canvas.
80 × 68 cm ($31\frac{1}{2} \times 26\frac{3}{4}$ in)
Old Reading Room, Ridgeway Library, The Library Company of Philadelphia.

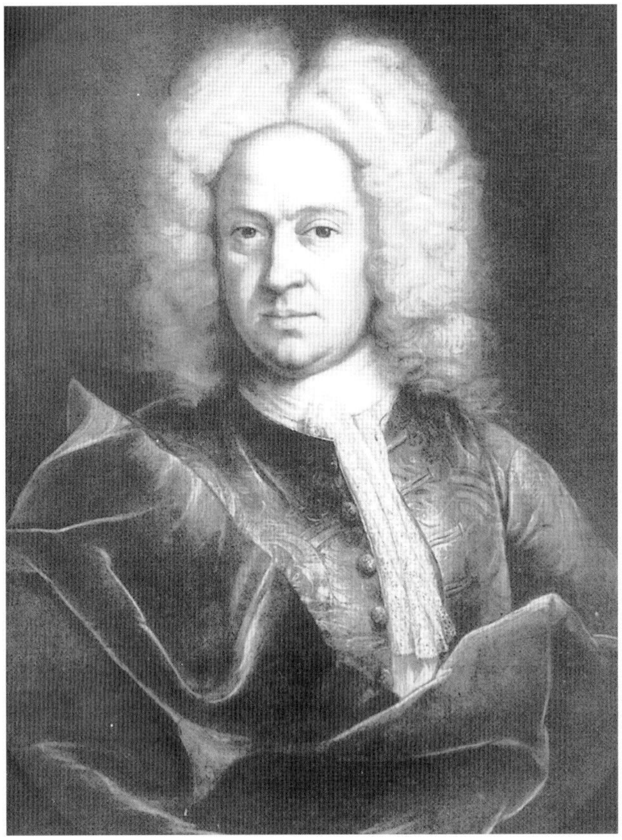

XXXIV-1

Half-length, with body half turned to right; head slightly to left, looking forward, with wig down to shoulders; white cravat tied at neck, with ends reaching halfway down chest, wearing unbuttoned, patterned coat, with loose robe covering right shoulder and arm.

James Logan (1674–1751) accompanied William Penn (1644–1718) to Pennsylvania in 1699 as his secretary. He assembled one of the finest libraries in the New World and imported the first copy of *Principia* (1687)[25] to reach America in April 1708, as well as owning a copy of *Opticks* (1704)[26] and, it is said, this portrait. Years after his death, the portrait was given in 1792 with the Loganian collection to the Library Company of Philadelphia, a subscription library started by Benjamin Franklin, FRS (1706–1790) in 1731. Despite this pedigree, the portrait bears no resemblance to Newton and looks Continental, *c.*1700.

XXXV. Lawrence (Lewis) Crosse (1650–1724)

XXXV-1 Portrait of bald man, oil on canvas.
Oval, 74 × 61 cm ($29\frac{1}{8} \times 24$ in)
The Burghley House Collection (no. 86), Stamford.

Half-length, man aged about fifty, with head half-turned to right, looking right; balding head, with reddish hair and brown eyes; steadying open book with left hand, and quill in right hand which is resting on book; wearing loose brown robe.

This is not a portrait of Newton, who 'was gray-headed when under 40',[32] and was never bald. His eyes were blue, and the features of the man in the portrait are not his. The portrait is hung with other oval paintings depicting members of the eighteenth-century Little Bedlam Club, of which Newton was never a member. In the first written reference to it in Drakard's Guide to Burghley House in 1815, the picture is labelled as Sir Isaac Newton by an unknown artist, but in a later nineteenth-century guidebook the artist is named as Lewis Crosse. There is a photograph of a portrait bust of the same unknown man in the Archive of the National Portrait Gallery also stated to be by Crosse.

XXXVI. Henry Cook (1642–1700)

XXXVI-1 Portrait, oil on canvas, 1669.
72.5 × 56.5 cm ($28\frac{1}{2} \times 22\frac{1}{4}$ in)
St Catharine's College, Cambridge.

Half-length, young man, with body turned to left; head half-turned to left, with eyes looking at viewer; long hair reaching to shoulders, with open shirt at neck with loose coat.

Signed and dated on the canvas, lower right.

In 1669, when he became Lucasian Professor, Newton was aged twenty-six. A number of pictures, that had been put away behind the College Library, were examined in 1943 or 1944, and one found 'much decayed which apparently represented a youngish man suffering from some form of madness'. On restoration and found to be signed by Cook and dated 1669, it was now thought to be a portrait of Newton, particularly when the oldest Fellow confirmed the belief that the College had indeed possessed such a portrait. Newton was a benefactor of St Catharine's College, though only long after the painting of the picture: at first he had loaned £25 in 1687, for which he was to receive an annuity of 50s. a year for life, but, as the annuity was never paid, he released the College from its obligation in 1714.[35] The portrait was said to resemble the Lely (**XXXVIII-1**) and John Riley (**XXXIX-1**) portraits, and not thought to be authentic by the Director of the National Portrait Gallery, Sir Henry Hake (1892–1951), a view accepted here.

XXXVII. Anthony Lee (*fl.*1724–1767)

XXXVII-1 Portrait, oil on canvas, 1747.
124.5 × 99 cm (49 × 39 in)
Private ownership.

Three-quarter-length, standing, facing half-left, in buttoned-up grey coat and tied white cravat at neck; wearing wig down to shoulders; left hand on hip, right hand half-slipped into coat; pillar with its base left side of canvas.

Signed, inscribed and dated 1747.

In Christie's sale of Great Tew Park, 28 May 1987 (lot 400). This painting is not accepted as a likeness of Sir Isaac Newton.

XXXVIII. Attributed to Sir Peter Lely (1618–1680)

XXXVIII-1 In January 1799, an engraving, 35.6 × 24.5 cm (14 × 9⅝ in), was published by Burnet Reading (1776–1822) with the inscription:

> Sir Isaac Newton when Bachelor of Arts in Trinity College, Cambridge. Engraved by Bt Reading from a head painted by Sir Peter Lely in the possession of the Right Honorable Lord Viscount Cremorne [1725–1813].

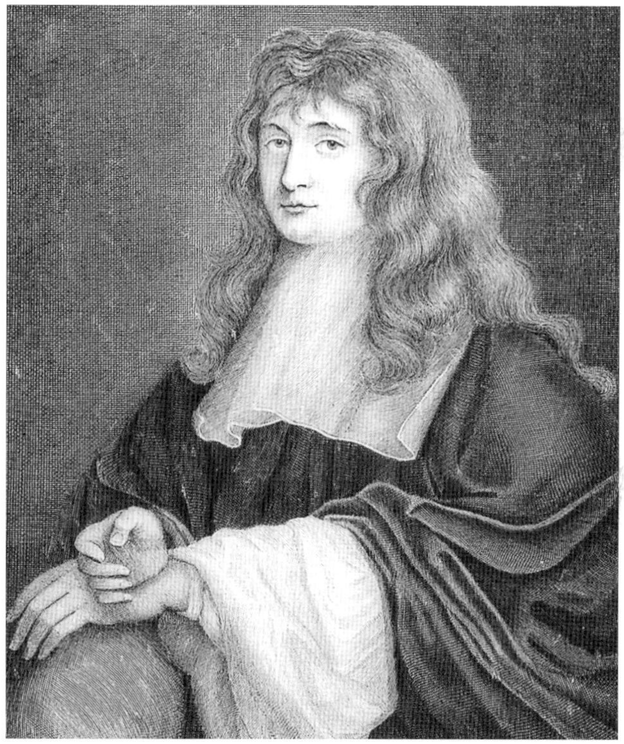

XXXVIII-1

The print shows an overweight young man, with neat long hair, in an academic gown, wearing very broad bands. Half-length, body turned half-left, with right hand across left wrist, which rests on a globe.

The features shown are not those of Newton, who was a Bachelor of Arts from 1665 to 1668. If an oil existed, it could not have been a portrait of Newton, and is most unlikely to have been painted by Lely, who died seven years before the publication of *Principia*[25] in

1687. There is no half-length portrait of Newton mentioned in R. B. Beckett or other studies of Lely,[3] or recorded in the Witt Library of the Courtauld Institute of Art.

XXXIX. John Riley (1646–1691)

XXXIX-1 Portrait, oil on canvas
76.2 × 63.5 cm (30 × 25 in)
With Ehrich, New York, 1929; sold in New York, 1 May 1930.

Half-length, head and body facing left, but with eyes looking towards viewer; wearing wig; short white cravat with loose outer garment.

This is not accepted as a likeness of Newton.

XL. Robert Sheppard (*fl.*1720–1740)

Gulliver's Travels by Jonathan Swift (1667–1745) was published in 1726, with a frontispiece engraving by Robert Sheppard. A.W. F. Edwards[13] has noted the similarity between the likeness of Captain Lemuel Gulliver in it and the portrait of Newton by Vanderbank, 1726 (**XIII-2**) in the Royal Society. There is indeed a similarity but insufficient to make it certain.

XLI. Unknown artist

XLI-1 Oil on canvas, late eighteenth century.
Oval, 73.5 × 61.5 cm (29 × 24¼ in)
Bonham's 22 September 1988 (lot 78) and 26 January 1989 (lot 85); Sotheby's 17 May 1989 (lot 22).

Half-length, body facing left, with head half-turned to his left and eyes to his right; untidy long white hair to below shoulder; wearing brown coat and white stock.

Inscribed upper left corner:
S^{R}. ISAAC NEWTON

This is not accepted as a likeness of Newton.

XLII. Christian Friedrich Zincke (1684?–1767)

Miniature enamel portrait.
Oval, 5 cm (2 in) high.
Earl of Gosford; Sotheby's 14 May 1925 (lot 304); Harry Seal; Christie's 16 February 1949 (lot 128); Sotheby's 13 December 1976 (lot 32).

Head and shoulders three-quarter to right; eyes looking at viewer, long grey hair, wearing a buff-coloured mantle over a blue coat, with tied white cravat.

This is not accepted as a likeness of Newton.

ENGRAVINGS

Knowledge of the engravings of Newton made before 1800 may be obtained from three main sources: F. O'Donoghue, *Catalogue of Engraved British Portraits Preserved in the Department of Prints and Drawings in the British Museum*, 1912,[40] D. E. Smith, 'Portraits of Sir Isaac Newton', 1927,[41] and R. Burgess, *Portraits of Doctors & Scientists in the Wellcome Institute*, 1973.[42] The prints have also been located in other collections, but particularly in the National Portrait Gallery, the Wren Library, Trinity College, and the Fitzwilliam Museum, Cambridge. These prints mainly derive from a limited number of the portraits in oil, and are here listed after the artists who painted them. The name of the engraver of a print, and particularly the date of its engraving, are not invariably discoverable: because of this and the great number of engraved portraits of Newton, it is unlikely that all those done in the eighteenth century have been included in this section, as well as likely that it contains engravings made after 1800.

Key to the source of the prints in this section:
RS = Royal Society
BM = Department of Prints and Drawings, British Museum
CUL = Cambridge University Library
DES = D. E. Smith Collection of Engravings, Columbia University[41]
FM = Fitzwilliam Museum, Cambridge
NPG = National Portrait Gallery
NMS = National Museum of Science and Industry
RAS = Royal Astronomical Society Archives
TCC = Trinity College, Cambridge
WL = Wellcome Library

XLIII. After Kneller, 1702 (III-1)

XLIII-1 Mezzotint, oval, 29.6 × 24.5 cm ($11\frac{5}{8} \times 9\frac{5}{8}$ in), 'Isaacus Newton Eq. Aur. G. Kneller Eques pinx. I. Smith Fecit et ex. 1712.' [John Smith (1652–1743)] [**BM**; **FM**; **RS**; **TCC**; **WL**]

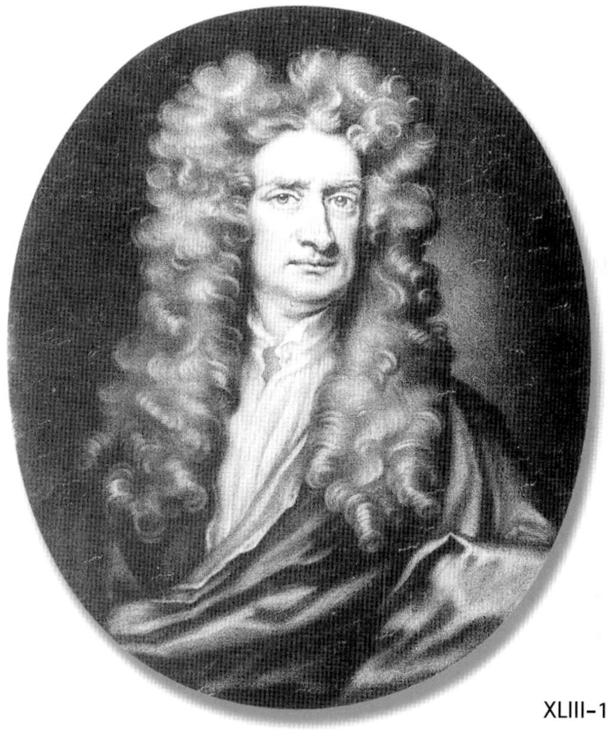

XLIII-1

XLIII-2 Line engraving, 36 × 22.7 cm ($14\frac{1}{4} \times 8\frac{7}{8}$ in): head in oval, 17 × 15.3 cm ($6\frac{7}{8} \times 6$ in) above pedestal with ornaments. 'Sr Isaac Newton. I. Houbraken Sculpsit. In the possession of John Conduit Esqr, 1743. G. Kneller pinx.' [Jacob Houbraken (1698–1780)] Plate in Thomas Birch, *Heads of Illustrious Persons of Great Britain*, 1743–52.[43] [Thomas Birch, FRS (1705-1766)] [**BM**; **FM**; **TCC**; **WL**]

John Conduitt was born in 1688 and died in 1737, and his wife, Newton's half-niece, Catherine Barton, lived from 1679 to 1740. So, in 1743 Kneller's portrait of

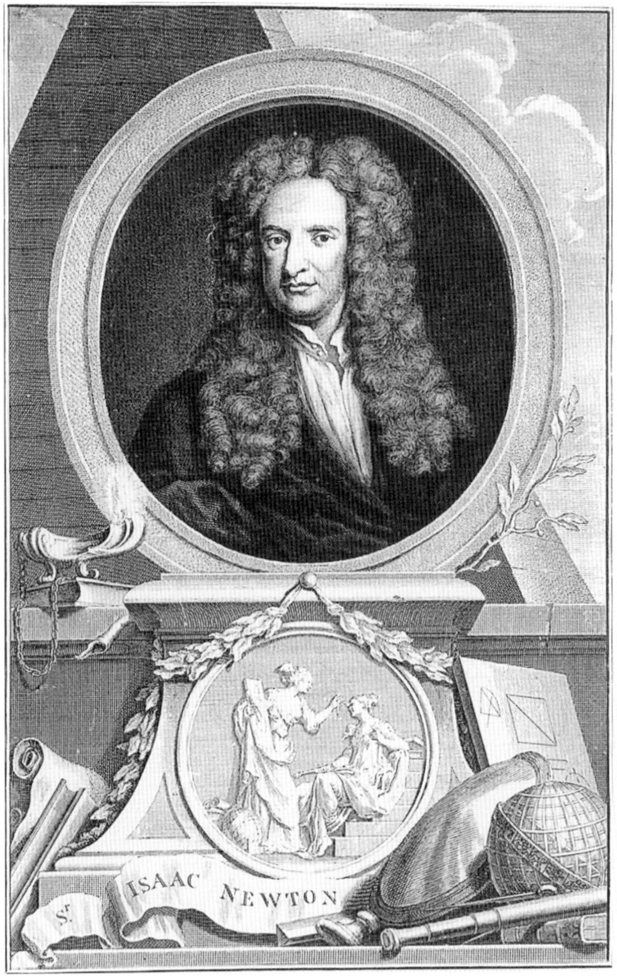

XLIII-2

1702 already belonged to their daughter Kitty (1718–1750), married, in 1740, to John Wallop (1718–1749), who became Viscount Lymington in 1743 when his father, also John Wallop (1690–1762), was created Earl of Portsmouth.

XLIII-3 Mezzotint, reversed copy from **XLIII-2**, with ornaments, engraver Andreas Miller (d.1763). Pub. P. Smith & A. Miller, Dublin, 1744. [**BM**]

XLIII-4 Line engraving, oval frame with ornaments, 16.5 × 10.2 cm ($6\frac{1}{2}$ × 4 in), 'S^{r}. Isaac Newton.' engraver not known. Plate on p. 288, *Universal Magazine*, vol. III, pub. J. Hinton, 1748. [**CUL**; **DES**; **WL**]

XLIII-5 Engraving, oval frame with ornaments, 15.2 × 11.1 cm (6 × $4\frac{3}{8}$ in) 'G. Kneller pinxt. Wm Sharp sculp^{t}. Sr Isaac Newton. Published by G. Kearsley, N° 46 Fleet Street.' [William Sharp (1749–1824)] [**BM**; **FM**; **RS**; **TCC**; **WL**]

XLIII-6 Engraving, oval in frame, 15.8 × 16.2 cm ($6\frac{1}{4}$ × $6\frac{3}{8}$ in) 'Sr Isaac Newton. G. Kneller pinx^{t}. J. Goldar Sculp. In the possession of John Conduit Esqr.' [John Goldar (1729–1795)] Engraved for M. (Paul) Rapin de Thoyras [1661–1725], *The History of England*. Translated from the French by N. Tindal (1687–1774). 6 vols. London: J. Harrison, 1785–9. [**DES**; **NPG**; **RS**; **TCC**; **WL**]

XLIII-7 Mezzotint, oval, 16.5 × 13.2 cm ($6\frac{1}{2}$ × $5\frac{3}{8}$ in), engraver John Simon (*c.*1675–1751). E. Cooper ex. [**BM**; **FM**; **NPG**; **TCC**]

XLIII-8 Mezzotint, reversed, oval, 16.5 × 13.2 cm ($6\frac{1}{2}$ × $5\frac{3}{8}$ in), engraved by J. & C. Esplens. [John and Charles Esplens (active *c.*1740)] [**BM**]

XLIII-9 Chalk manner engraving, printed within decorated border, engraved by J. C. François. [Jean Charles François (1717–1769)] Plate in Saverien's *Hist. des Philosophes Modernes*, 1760. [**BM**]

XLIII-10 Engraving, 12.7 × 8.1 cm (5 × $3\frac{1}{4}$ in), 'Isaac Newton. né en 1643. mort en 1726. J. Blanchoir in. et sculp^{t}.' [**DES**]

XLIII-11 Engraving, one of four heads (no. 3) done from different portraits of Newton, oval, 10.9 × 8.7 cm ($4\frac{1}{4}$ × $3\frac{7}{16}$ in). 'Joh: H. Lips fec: 1778'. [Johann Heinrich Lips (1758–1817)] Plate XII, vol. IV, in Johann Kaspar Lavater [1741–1801], *Physiognomische Fragmente*. 4 vols. Leipzig, 1775–8. [**CUL**; **TCC**] – see p. 65.

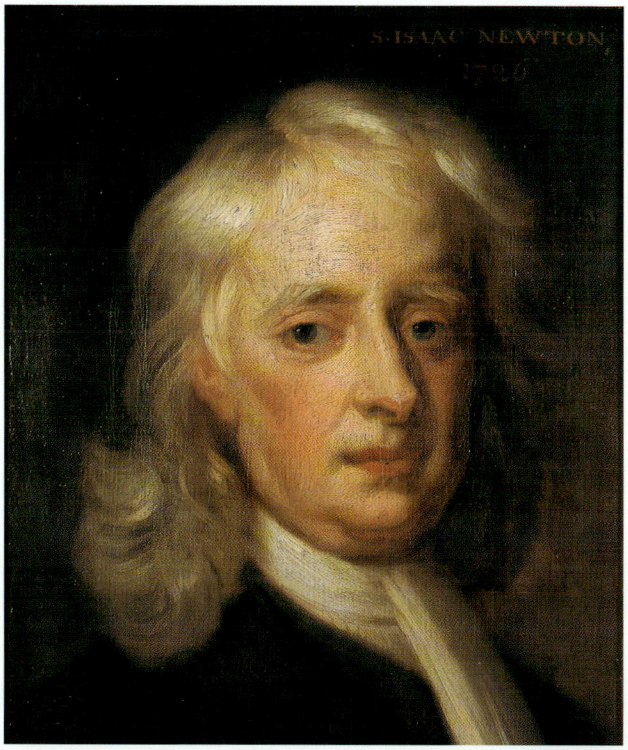

LEFT: **XIV-1** Head and shoulders of Newton, oil on canvas sketch, by Enoch Seeman, 1726.

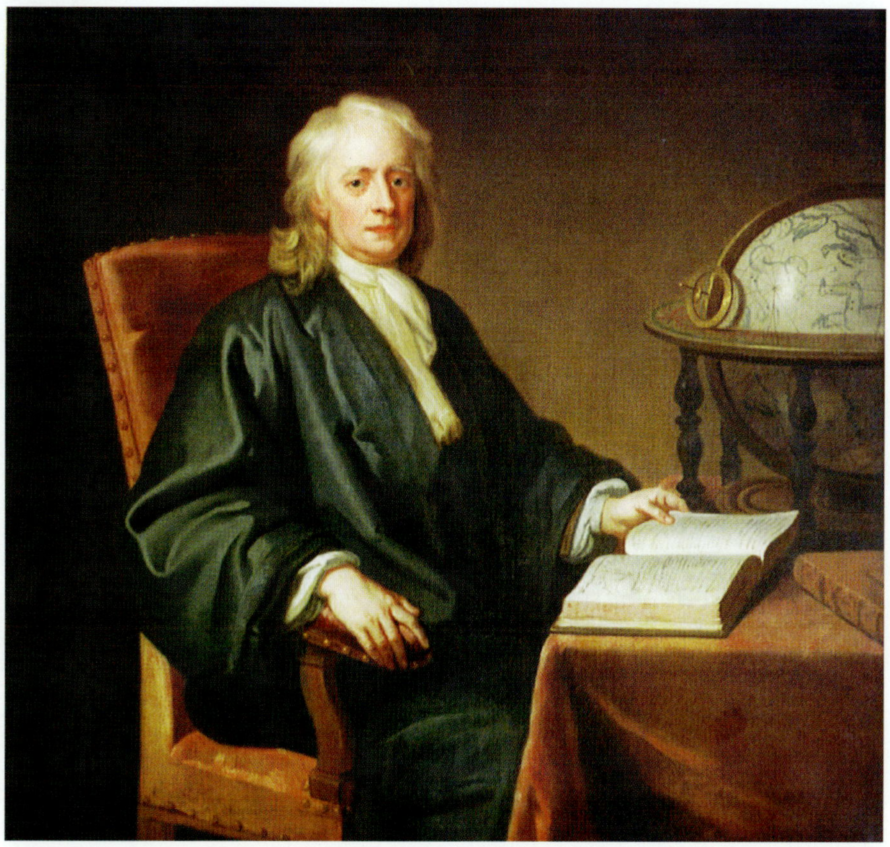

RIGHT: **XIV-2** Newton aged eighty-three, oil on canvas, by Enoch Seeman, 1726.

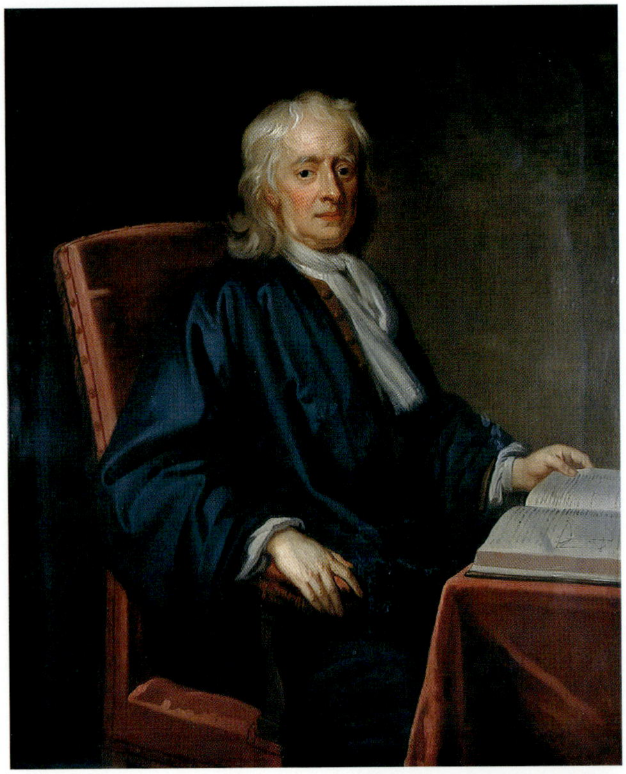

LEFT: **XIV-3** Newton, oil on canvas, after **XIV-2**, by Enoch Seeman.

BELOW: **XIV-4** Newton, oil on canvas, after **XIV-2**, by unknown artist.

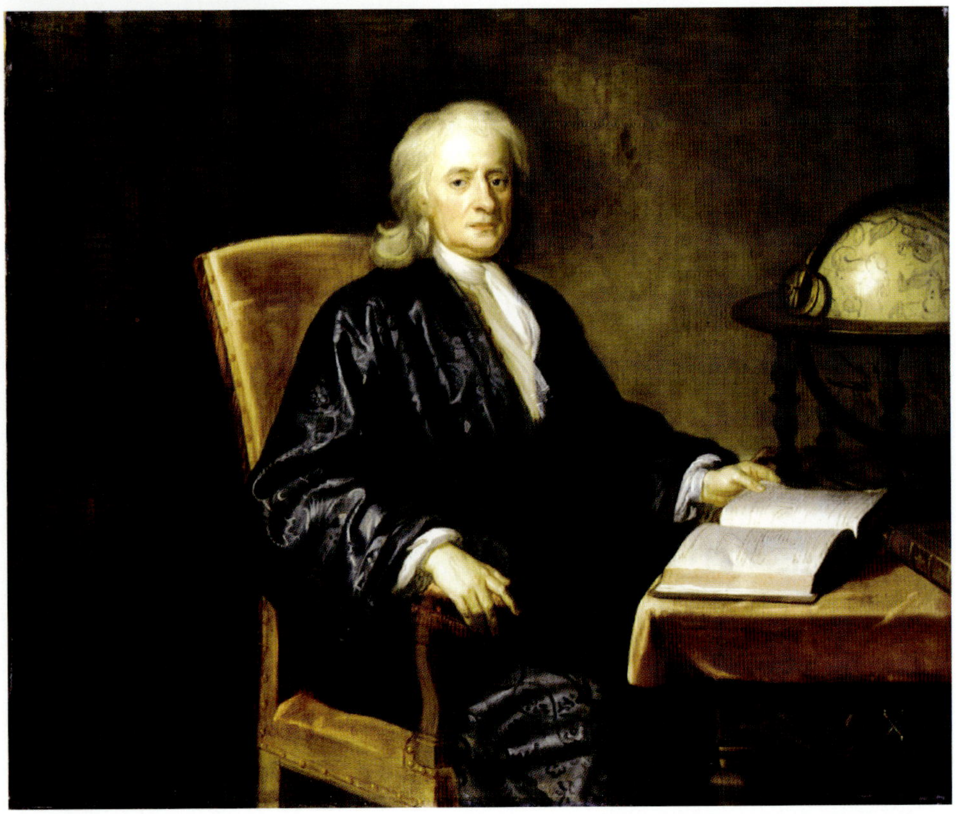

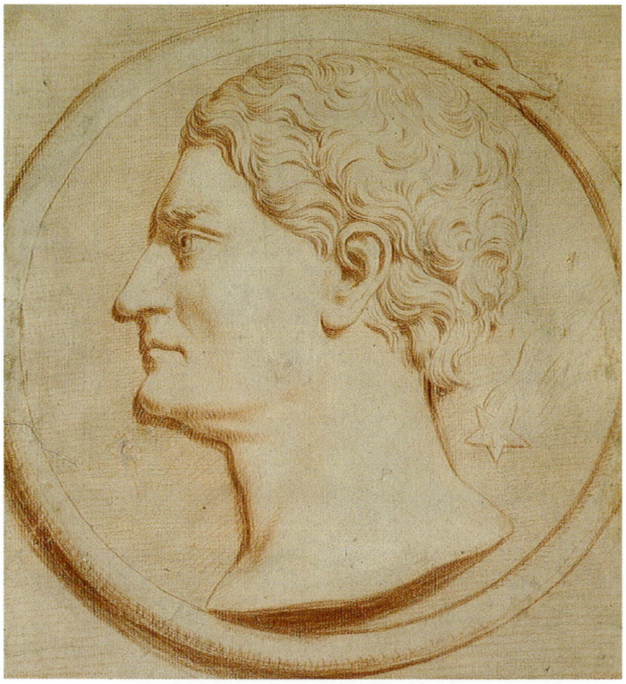

ABOVE: **XVII-1** Drawing of Newton in red chalk on paper by William Hoare, 1734.

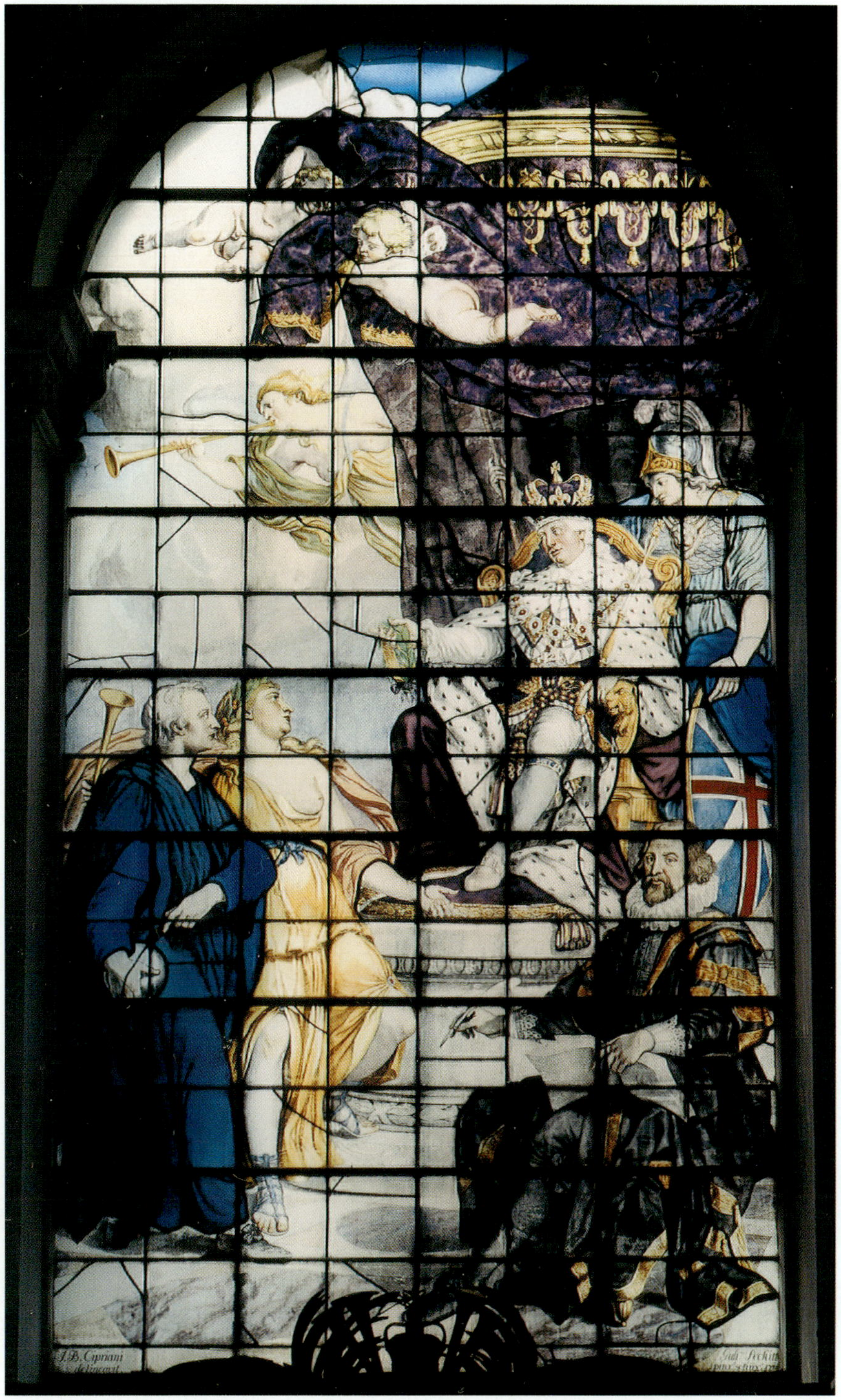

RIGHT: **XXIV-4** Coloured glass window in Wren Library, Trinity College, Cambridge, 1775.

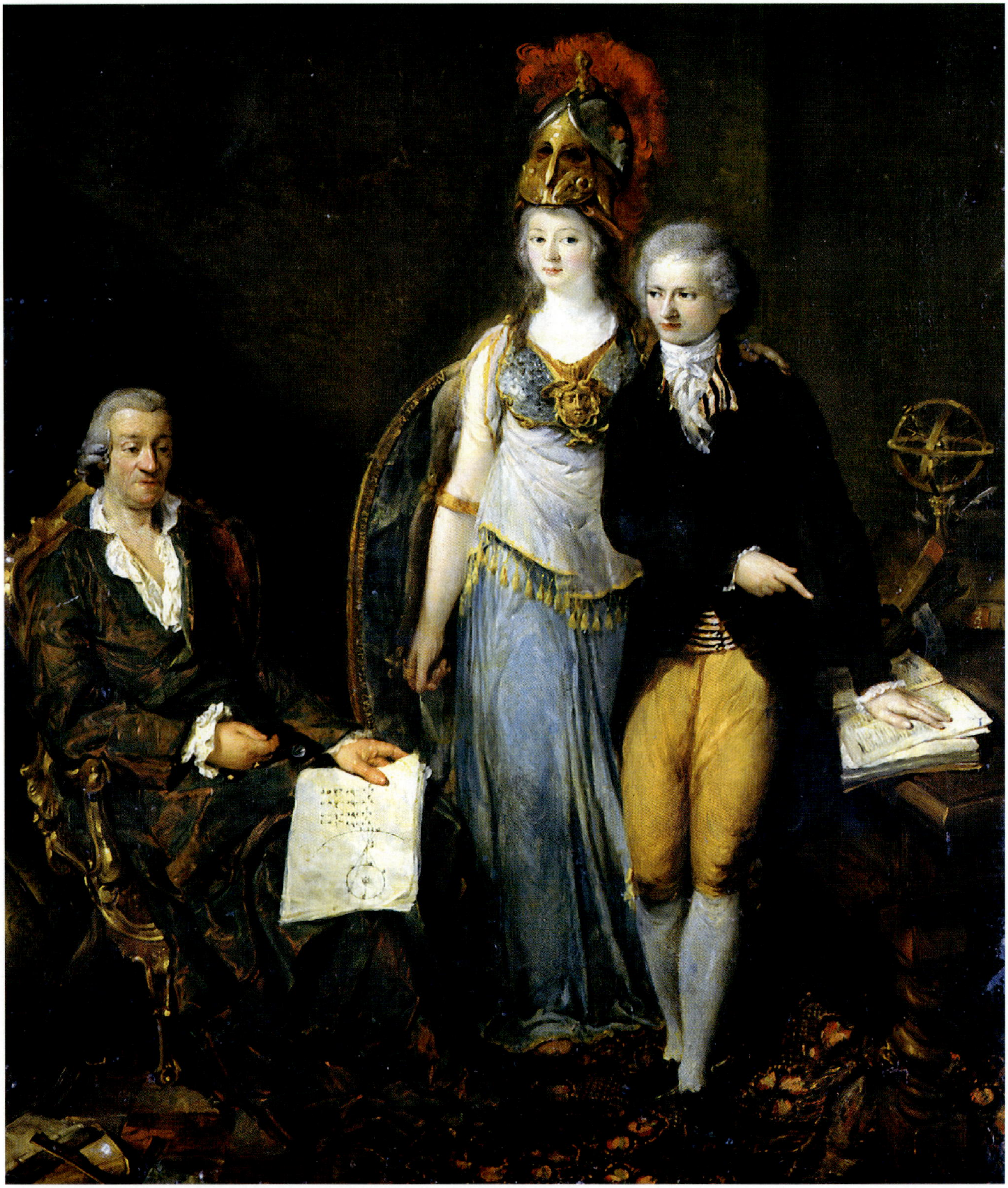

XXIX-1 Allegorical painting *Britannia between Newton and W. J. 'sGravesande*, oil on canvas, by G. M. de Boni, 1787.

RIGHT: **B.1** Relief of Newton in ivory plaque by David le Marchand, 1702–10.

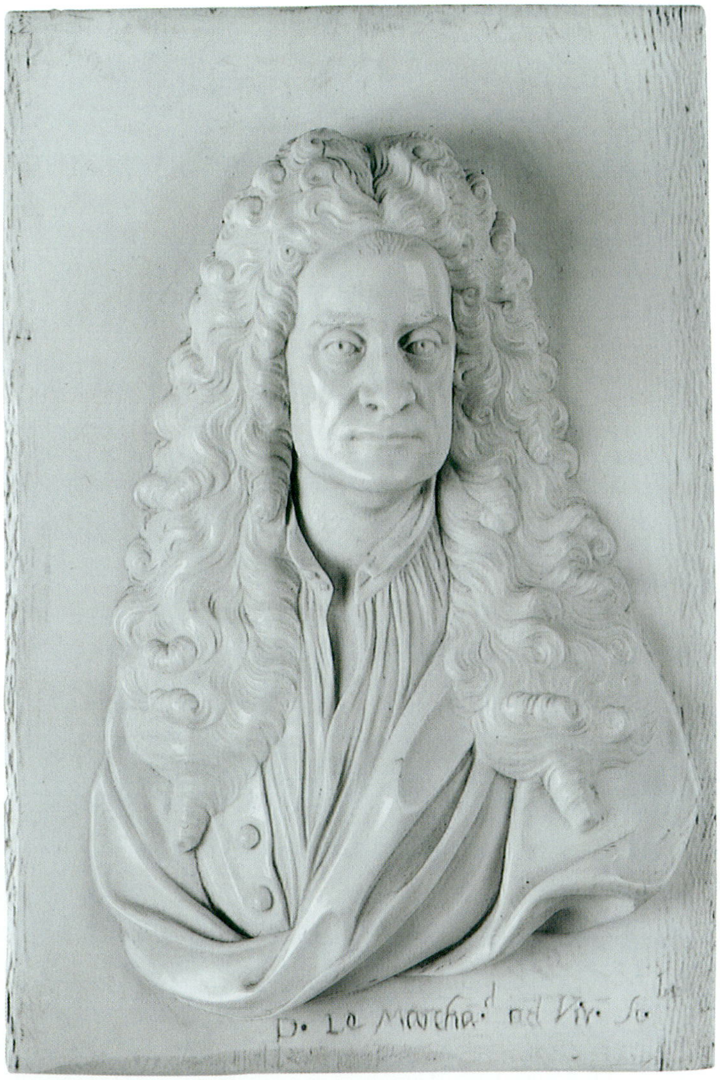

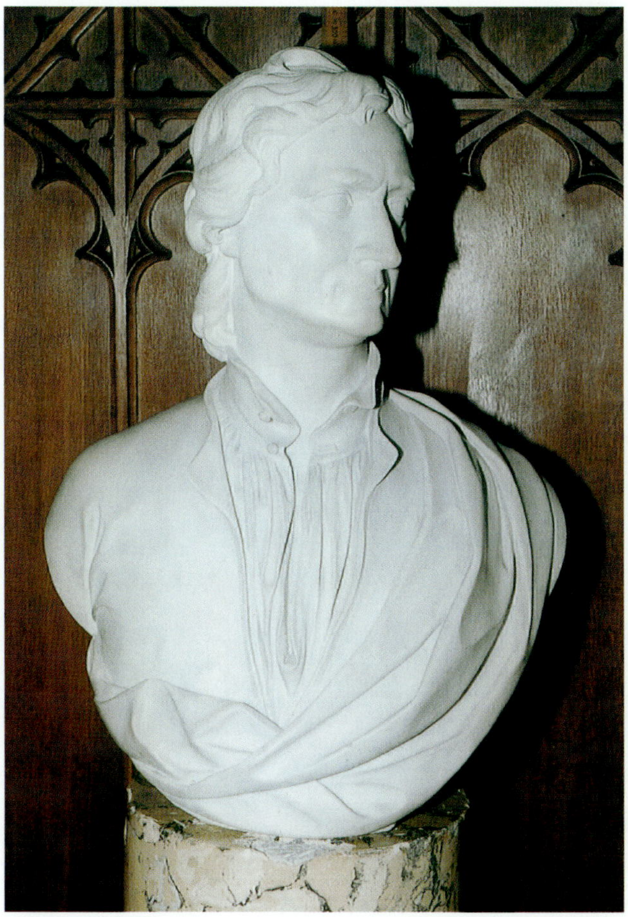

LEFT: **A.8** Stone bust of Newton by G. B. Guelfi, c.1731, at Scone Palace, Perthshire.

RIGHT: **A.10** Marble bust of Newton, the 'Conduitt bust', by Michael Rysbrack, c.1727–29.

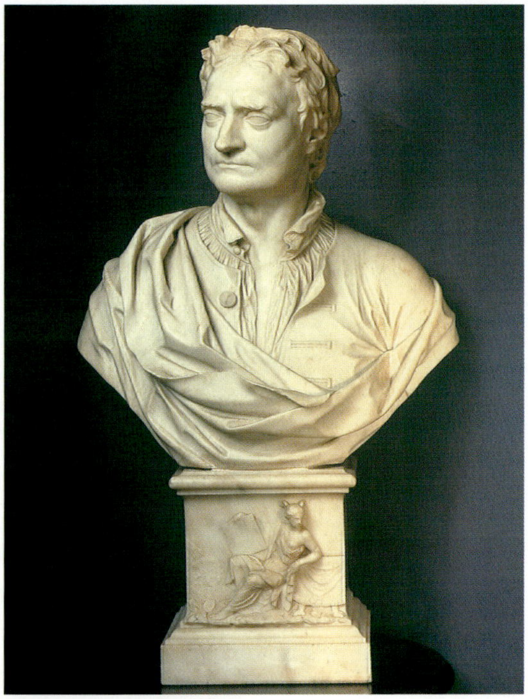

BELOW: **A.11** Bust of Newton in *A scene from 'The Indian Emperour' or The Conquest of Mexico by the Spaniards*, oil on canvas, by William Hogarth, 1732–35.

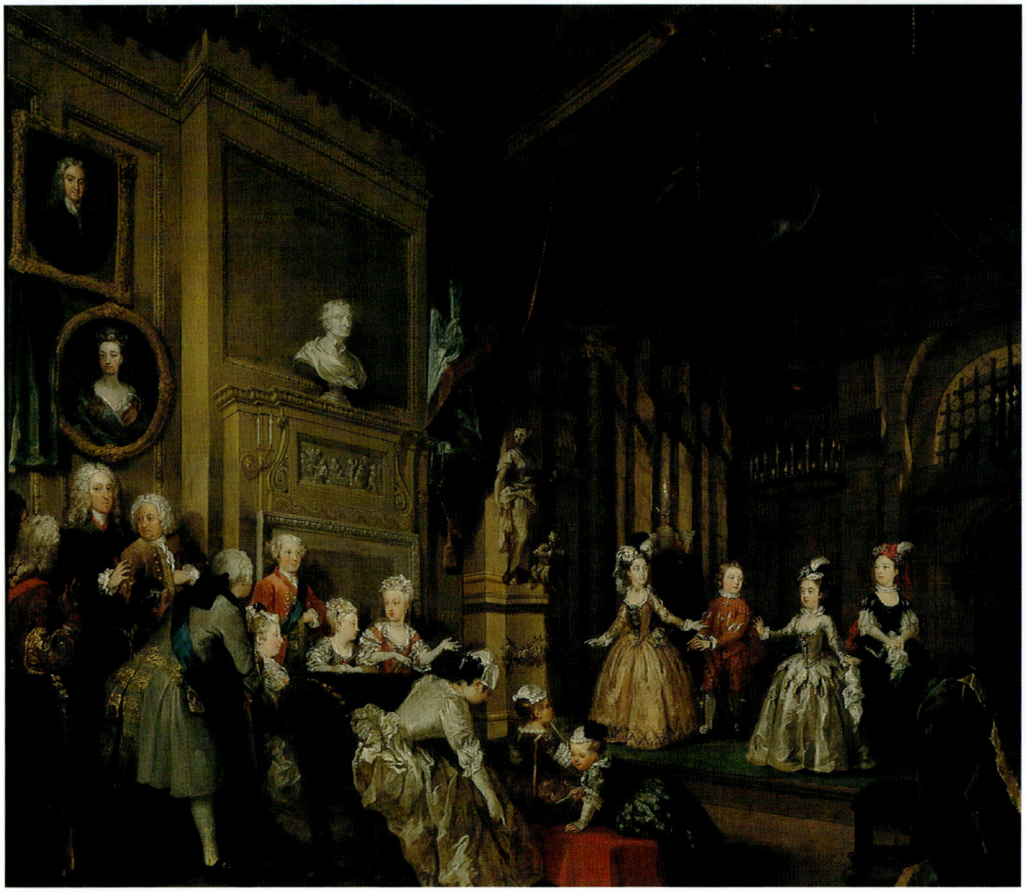

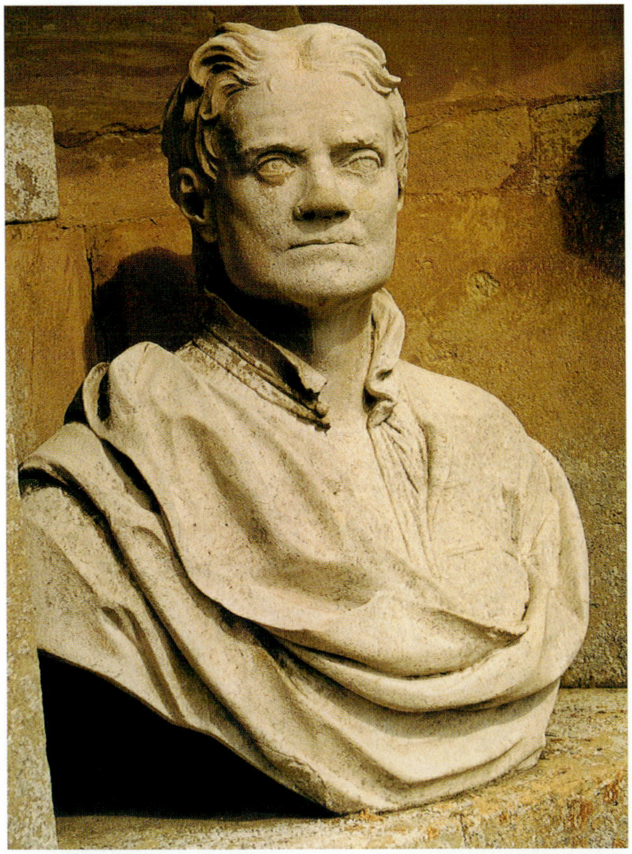

LEFT & BELOW: **A.12** Left, stone bust of Newton by Michael Rysbrack, *c.*1727, in Temple of British Worthies at Stowe, Buckinghamshire. Below, the bust on Newton's right is of John Locke and to his left that of Sir Francis Bacon, both also carved by Rysbrack.

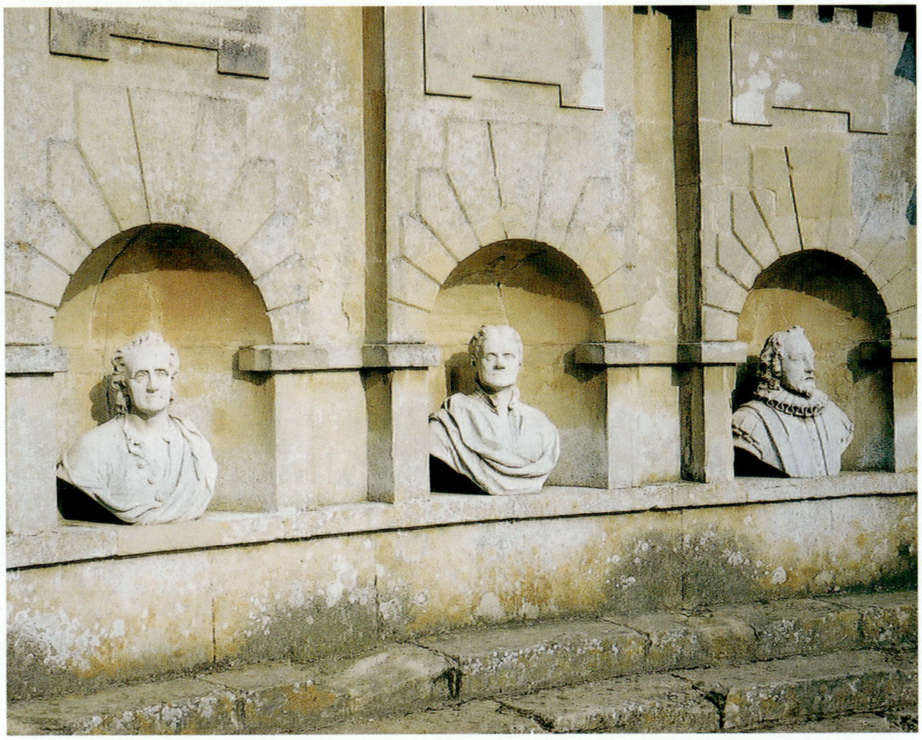

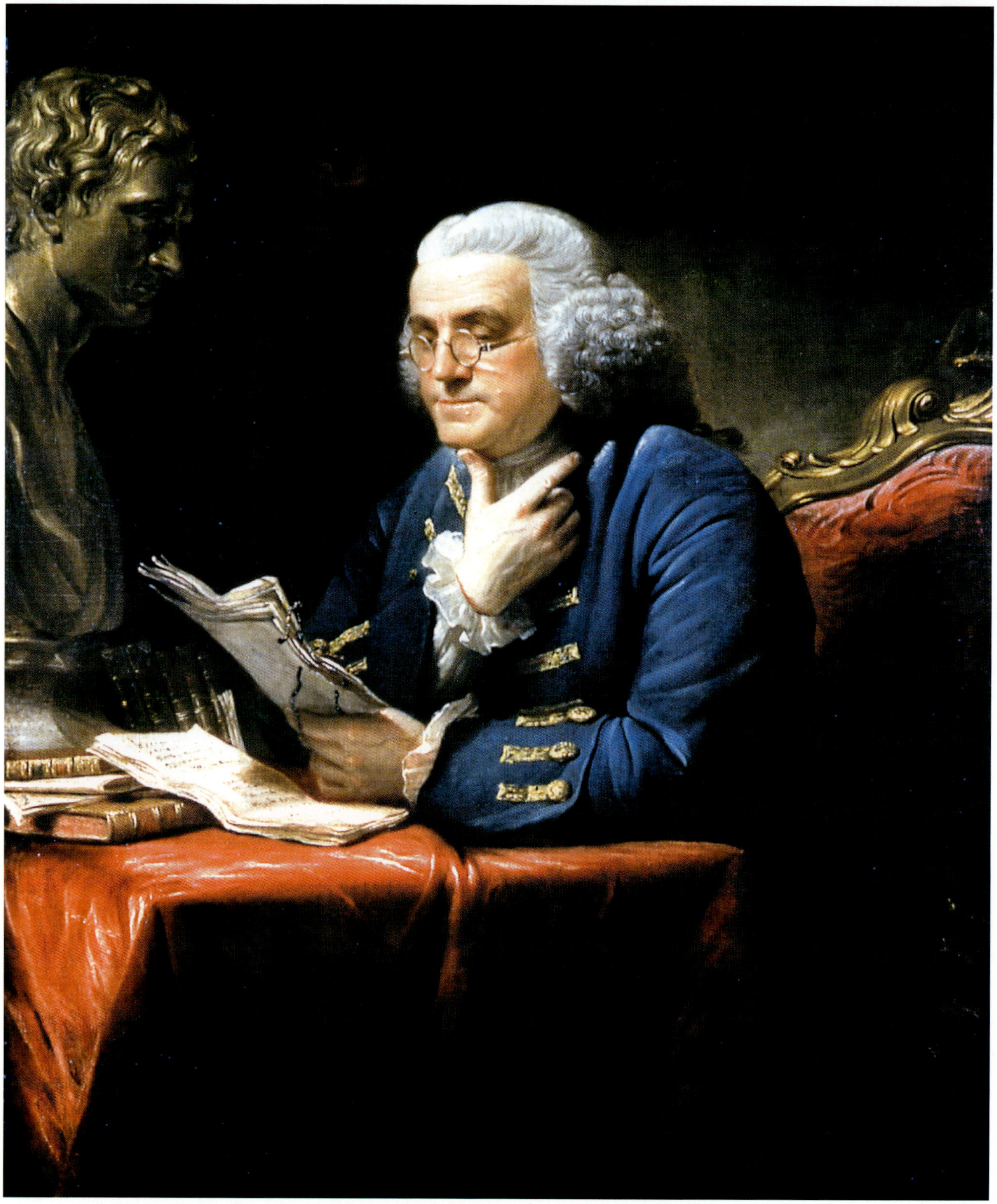

A.17 A bust of Newton in portrait of *Benjamin Franklin*, oil on canvas, by David Martin, 1766.

XLIII-12. Engraving, oval, 18.1 × 11.6 cm ($7\frac{1}{8} \times 4\frac{9}{16}$ in), 'des Rais del. Ex Biblioth^{ca}. Regia. Le Coeur Sculp^{t}. Newton.' [Louis Le Coeur (*fl.*1780– after 1800)] [**DES**]

XLIII-13 Line engraving, oval within rectangle, 17.3 × 12 cm ($6\frac{3}{4} \times 4\frac{3}{4}$ in), 'Natus A.D. 1642. Sr. Isaac Newton. Denatus A.D. 1726.', engraver not known. [**TCC**; **WL**]

XLIV. After William Gandy, Junior, 1706 (XXXIII-1)

XLIV-1 Engraved lithograph, 31.7 × 27 cm ($12\frac{1}{2} \times 10\frac{5}{8}$ in), 'AEtatis Suae 64. 1706. Drawn on Stone by G. B. Black, 1848. Painted by William Gandy, Junr 1706. Day & Son, Lith^{rs} to the Queen. Sir Isaac Newton, From an original painting 30 in. by 26 in. formerly in the possession of John Abel Walter, Esq. of Ashley Mount, near Lymington. Published by R. A. Grove, Lymington, 1848.' [**BM**; **NPG**; **DES**; **WL**]

No likeness to Newton can be discerned in this lithograph.

XLIV-2 Engraving of head and shoulders, 9 × 8 cm ($3\frac{1}{2} \times 3\frac{1}{8}$ in) [**RS**]

XLV. After Thornhill (IV-2)

XLV-1 Mezzotint, half-length figure, 30.5 × 24.8 cm ($12 \times 9\frac{3}{4}$ in), 'Isaacus Newton Eq. Aura. I Thornhill Eq. pinx I Simon fec. Sold by I Simon in New Street Covt Garden.' [Jean Simon (*c.*1675–1751)] [**FM**; **TCC**; **WL**]

Mentioned in a letter from Roger Cotes to Newton on 20 July 1712.[27]

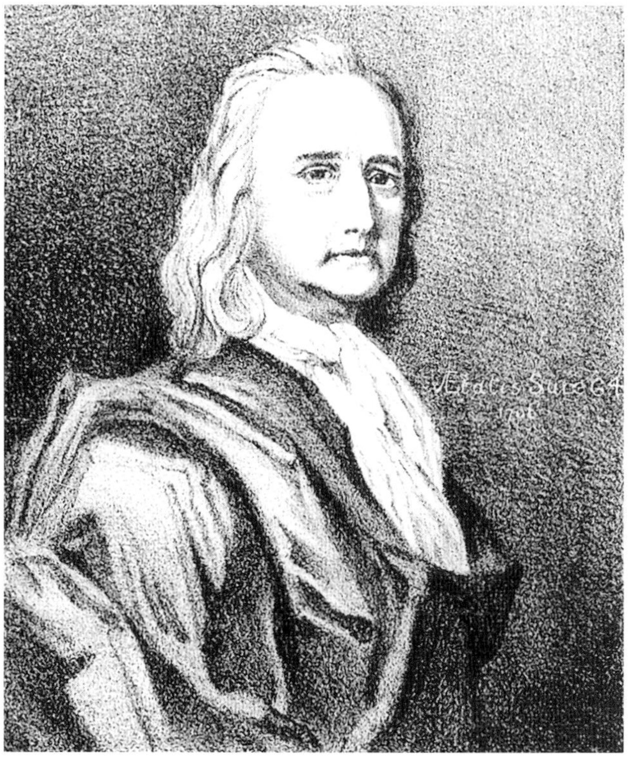

XLIV-1

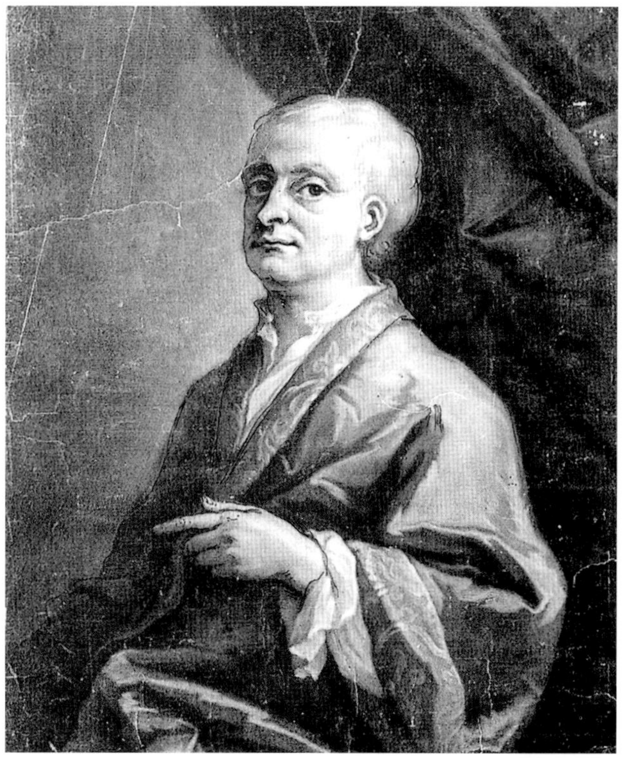

XLV-1

XLVI. After Richter, 1714 (V-1)

XLVI-1 Engraving, medallion, 12.5 × 8.8 cm ($4\frac{7}{8} \times 3\frac{1}{2}$ in), 'Richter pinx. 1714. SIR ISAAC NEWTON. from a Miniature picture of the same size. London, Publish'd by C. Taylor. N°10 … Street, Holborn, Novr 1, 1785.' [**DES**]

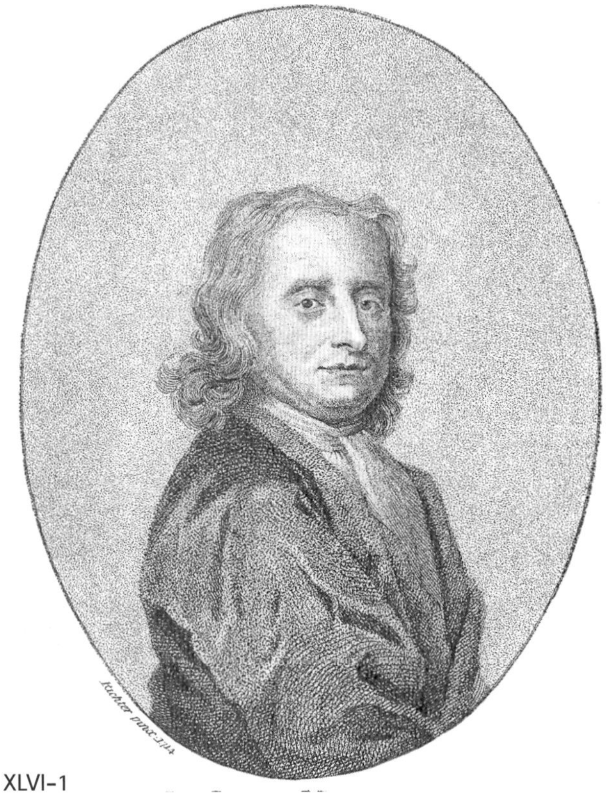

XLVI-1

XLVI-2 Engraving, oval, stipple, 10 × 7.6 cm (4 × 3 in), 'Richter pinx. 1714. SIR ISAAC NEWTON.' Pub. W. Smith, 1814. [**BM**]

XLVII. After Vanderbank, 1726 (XIII-3)

XLVII-1 Mezzotint, 32 × 24.7 cm ($12\frac{5}{8} \times 9\frac{3}{4}$ in), 'Isaacus Newton Equ^{s}. Anno AEtat 84. A C 1726. J. Vanderbank pinx 1725. J. Faber Fecit 1726. J. Cooper Excudit.' [John Faber, the younger (1684–1756)] – various impressions with re-engraved inscriptions. Mezzotint bequeathed to Fitzwilliam Museum by Richard, 7th Viscount Fitzwilliam (1745–1816), in 1816. [**BM**; **FM**; **NPG**; **RS**; **TCC**; **WL**]

XLVII-2 Line engraving, oval, 5 × 3.8 cm ($2 \times 1\frac{1}{2}$ in), 'Vanderbank p^{t}. A. Smith, sc. Pub^{d} by Harrison & Co April 1, 1794. Sir Isaac Newton.' [Anker Smith (1759–1819)]. Plate in *Biographical Magazine*. [**BM**; **DES**; **WIL**]

XLVII-3 Engraving, one of four heads (no. 2) done from different portraits of Newton, oval, 10.9 × 8.7 cm ($4\frac{1}{4} \times 3\frac{7}{16}$ in). 'Joh: H. Lips fec: 1778'. [Johann Heinrich Lips (1758–1817)] Plate XII, vol. IV, in Johann Kaspar Lavater [1741–1801], *Physiognomische Fragmente*. 4 vols. Leipzig, 1775–8. [**CUL**; **TCC**] – see p. 65.

XLVII-4 Line engraving, 9.8 × 5.4 cm ($3\frac{7}{8} \times 2\frac{1}{8}$ in), bust in oval with ornaments, 'Delamonce del A Duflos scul' [Ferdinand Delamonce (1678–1753); Claude-Augustin-Pierre Duflos (1700–1786)] Engraving on title page of J. Castillioneus, *Isaaci Newtoni… OPUSCULA*, 1744.[44] [**TCC**; **DES**]

XLVII-5 Engraving, 17.6 × 11.2 cm ($6\frac{7}{8} \times 4\frac{3}{8}$ in), 'Literary Magazine & British Review. S^{R} Isaac Newton. Published as the Act directs 1. Decr 1788 by C. Forster No 41 Poultry.' [**DES**]

XLVII-6 Engraving, oval, 8.3 × 7.9 cm ($3\frac{1}{4} \times 3\frac{1}{8}$ in), 'R. Scott Sc. Sir Isaac Newton.' [Robert Scott (1771–1841)] [**DES**]

XLVII-7 Engraving, head in oval, 9.2 × 7.2 cm ($3\frac{5}{8} \times 2\frac{7}{8}$ in), 'S^{r}. Isaac Newton.' [**TCC**]

XLVII-8 Engraving, head in oval, 9.8 × 7.9 cm ($3\frac{7}{8} \times 3\frac{1}{8}$ in), 'S^{R}. ISAAC NEWTON.' [**TCC**]

XLVIII. After Vertue, 1726 (XIII-4)

XLVIII-1 Line engraving, 18.7 × 15 cm ($7\frac{3}{8} \times 5\frac{7}{8}$ in), 'Isaacus Newton Eq. Aur. AEt. 83. I. Vanderbank pinxit 1725. Geo. Vertue Sculpsit 1726.', frontispiece of third edition of *Principia*,[25] 1726. [**TCC**]

XLIX. After Seeman, 1726 (XIV-1)

XLIX-1 Mezzotint, 28 × 18.5 cm ($11 \times 7\frac{1}{4}$ in), 'Sir Isaac Newton. Drawn and Scraped MDCCLX by James Macardel [James MacArdell (1710?–1765)] from an original portrait painted by Enoch Seeman now in the possession of Thomas Hollis F. R. and A. S. S.' Below this a quotation of several lines from Voltaire: 'Les Italiens, ces peuples in Génievx ont craint de penser, les François n'ont osé penser … Ode sur la mort de Madame de Bareith avec une lettre par Mons. de Voltaire.' Plate in *Memoirs of Thomas Hollis Esq.*, 2 vols., privately printed, 1780, vol. II, p. 800. [**NPG**; **TCC**; **WL**]

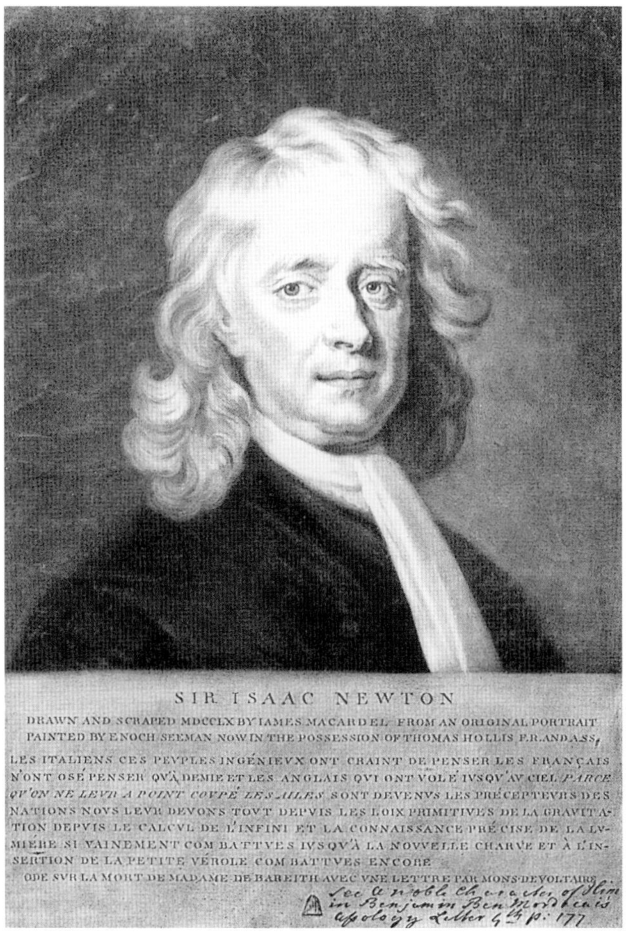

XLIX-1

XLIX-2 Engraving, one of four heads (no. 1) done from different portraits of Newton, oval, 10.9 × 8.7 cm ($4\frac{1}{4} \times 3\frac{7}{16}$ in). 'Joh: H. Lips fec: 1778'. [Johann Heinrich Lips (1758–1817)] Plate XII, vol. IV, in Johann Kaspar Lavater [1741–1801], *Physiognomische Fragmente.* 4 vols. Leipzig, 1775–8. [**CUL**; **TCC**] – see p. 65.

XLIX-3 Engraving, 32.2 × 15.5 cm ($12\frac{5}{8} \times 6\frac{1}{8}$ in), 'E. Seeman, p^{t}. Monsaldi Sc.' [Antoine Maxime Monsaldi (1768–1816)] [**DES**; **WL**]

XLIX-4 Engraving, 22.5 × 16 cm ($8\frac{7}{8} \times 6\frac{1}{4}$ in), 'J. Swaine. Sculp. Sir Isaac Newton. From an original by Enoch Seeman. *another of the managers Js Newton. Js Newton.* His autographs from the Originals in the possession of John Thane.' [John Swaine (1775–1860)] [**DES**; **WL**]

XLIX-5 Line engravings, two portraits in ovals on same plate: above, **after Seeman, 1726 (XIV-2)**, 15.2 × 11.3 cm ($6 \times 4\frac{1}{2}$ in); below, **after Vanderbank, 1726 (XIII-2)**, 10.3 × 8.4 cm ($4\frac{1}{8} \times 3\frac{5}{8}$ in). [**DES**; **TCC**]

L. After Vanderbank, 1727 (XV-1)

L-1 Mezzotint, 35.4 × 25.4 cm ($12\frac{7}{8} \times 10$ in), 'Isaacus Newton Equ^{S}. Obijt 20 Mar. 1726/7. AEtatis Suae 85, I. Vanderbank pinx. An^{O} 1726 Pro Martino Folkes Arm^{O} Cui hanc Tabulam D. D. I. Faber.', engraver John Faber, the younger (1684–1756). [**DES**; **WL**]

LI. After Pittoni with D. and G. Valeriani, 1730 (D.9)

LI-1 Engraving of painting *An allegorical monument to Sir Isaac Newton*, with arched top, 60 × 39.4 cm ($23\frac{5}{8}$ × $15\frac{1}{2}$ in).
Inscribed: 'J. B. Pittoni et D. et Jos. Valeriani Pinx. L. Desplaces Sculp. et D. M. Fratta delin. / Vivida vis animi pervicit, & extra / Processit longe flammantia Moenia Mundi / Atque omne immensum peragravit Mente Animoque (Lucretius lib. I)' ('The triumphant power of his mind went far beyond the fiery confines of the world and ranged through the whole of creation with keen intelligence.') [Louis Desplaces (1682–1739); Domenico Maria Fratta (1696–1763)]

Plate XIX in Owen (Eugene) MacSwinny [(before 1684–1754)], *Tombeaux des princes, des grands capitaines et autres hommes illustres qui ont fleuri dans la Grands-Bretagne vers la fin du XVII et le commencement du XVIII siècle … Le tout dirigé et mis au jour, par les soins de E.M.* Londres, 1741. In the event only nine plates of the twenty-four paintings of imaginary monuments, that were commissioned by MacSwinny from Venetian and Bolognese artists in Venice between 1720 and 1730, were engraved.
[**BM**; **FM exhibition catalogue 1974**;[10] **TCC**; **WL**]

LII. After François Boucher (1703–1770)

LII-1 Line engraving and etching, '*Allégorie*', 1741, 16.2 × 32 cm ($6\frac{3}{8}$ × $12\frac{1}{2}$ in), inscribed 'F. Boucher invenit et delin. Lau: Cars sculpsit.' [Laurent Cars (1699–1771)].

Half-title, with medallion of Newton crowned with a laurel wreath and inserted into a pyramid, with the signs of the Zodiac behind, and two putti, a bull, a lobster, a lion and a ram in foreground, preceding 'Abregé de l'Éloge de M. le Chevalier Newton par M. de Fontenelle' in Owen MacSwinny, *Tombeaux des princes, des grands capitaines et autres hommes illustres qui ont fleuri dans la Grands-Bretagne vers la fin du XVII* … Londres, 1741. The pages on Newton start with a title page designed by Pierre Jos. Perrot (*fl.*1724–1835) and engraved by Cars with a border in Louis XIV style, and finish with another half-title, also designed by Boucher and engraved by Cars, which has a book with an arrangement of scientific instruments, including an armillary sphere.[10] [**BM**; **TCC**; **RAS**; **WL**]

LIII. After Kent and Rysbrack, 1731 (D.6)

LIII-1 Line engraving of Monument erected to the Memory of Sir Isaac Newton in Westminster Abbey, 28 × 21.2 cm (11 × $8\frac{3}{8}$ in), 'Ryley del. published by R. Bowyer, Historic Gallery, Pall Mall, May 20 1794. Fittler Sculp.' [Charles Reuben Ryley (1752–1798); James Fittler (1758–1835)] [**TCC**; **WL**]

LIV. After Guelfi, 1732 (A.9)

LIV-1 Mezzotint, 31.3 × 24.8 cm ($12\frac{1}{2} \times 9\frac{3}{4}$ in), marble bust in a niche, 'S^R. ISAAC NEWTON. Done after the Marble Bust in Her Majesty's Hermitage in the Royal Garden at Richmond. *Printed for Ca Bowles in St Pauls Church Yard, London*'. Engraved by John Faber, the younger (1684–1756) *c.*1736. One of a set of plates from busts of philosophers in Queen Caroline's Grotto at Richmond. ⌜**NPG**; **WL**⌝

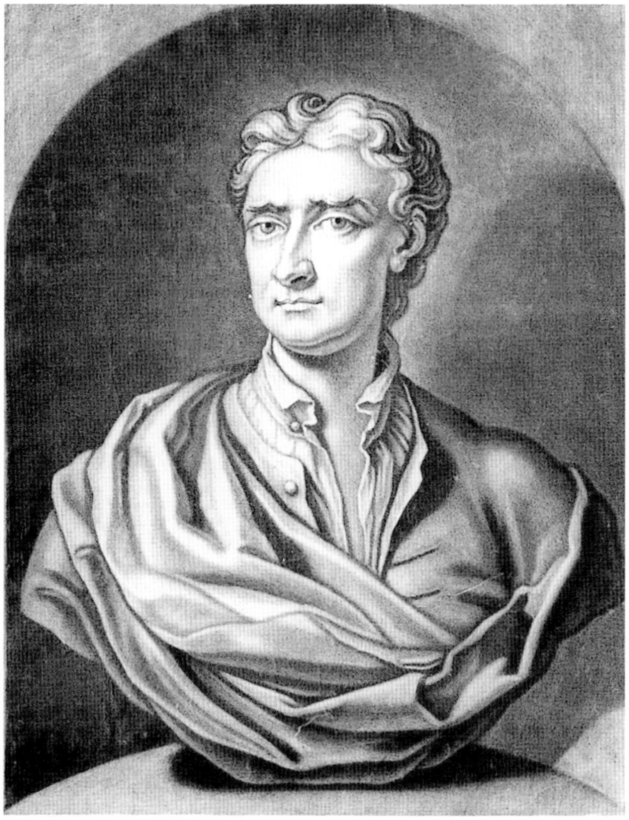

LIV–1

LIV-2 Engraving, one of four heads (no. 4) done from different portraits of Newton, oval, 10.9 × 8.7 cm ($4\frac{1}{4} \times 3\frac{7}{16}$ in). 'Joh: H. Lips fec: 1778'. ⌜Johann Heinrich Lips (1758–1817)⌝ Plate XII, vol. IV, in Johann Kaspar Lavater ⌜1741–1801⌝, *Physiognomische Fragmente.* 4 vols. Leipzig, 1775–8. ⌜**CUL**; **TCC**⌝ – see p. 65.

LV. George Bickham, Senior (1684–1769), 1732

LV-1 Engraving, bust-profile in circle surrounded by geometric rays in centre of starry firmament, with allegorical figures, putti and scientific instruments, and laudatory verse below, 34 × 24 cm ($13\frac{3}{8} \times 9\frac{1}{2}$ in). 'G. Bickham, Senior, Sculpt., 1732. Sold by G. Bickham Snr Strand Coffee House, nr Catharine St. …' New impression sold by John Bickham, engraver, in 1752, and another: *London Published May 8th 1787 by Robert Sayer 53 Fleet Street.* ⌜**NMS**; **RAS**; **RS**; **WL**⌝

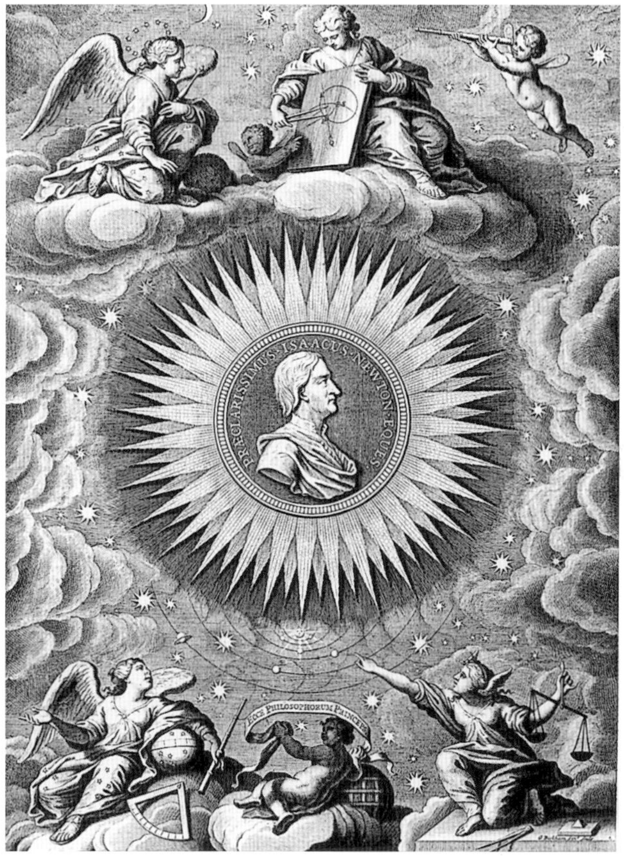

LV–1

LVI. Jacob Smith (*fl.*1730), 1733

LVI-1 Line engraving, 20 × 34.2 cm ($7\frac{7}{8} \times 13\frac{1}{2}$ in), a head of Newton with another of Sir Hans Sloane; two circles executed with one spiral line, 1733, engraved by Jacob Smith. [**BM**]

LVII. After William Hoare, 1734 (XVII-1)

LVII-1 Etching, profile head of Newton in the eternal circle of a serpent with its tail in its mouth, the *ouroborus*, 29 × 27.6 cm ($11\frac{3}{8} \times 10\frac{7}{8}$ in), etched by William Hoare, 1734. [**WL**]

LVIII. After Louis-Fabricius Dubourg, 1738 (XVIII-1)

LVIII-1 Engraving, 14 × 9 cm ($5\frac{1}{2} \times 3\frac{1}{2}$ in), 'L. F. Dubourg inv. I. Folkema sculp'. Frontispiece of F. M. Arouet de Voltaire (1694–1778), *Elémens de la Philosophie DE NEUTON* par M^{R}. DE VOLTAIRE, 1738.[34] [Jacob Folkema (1692–1767)] [**CUL**; **TCC**]

LIX. After Jacques Roëttiers (1707–1784),[46] 1739 (G.4)

LIX-1 Engraving, profile head in oval, 12.4 × 6.1 cm ($4\frac{7}{8} \times 2\frac{3}{8}$ in), 'Isaac Neuton. Mort à Londres, le 20. Mars 1727. Agé de 85 Ans. Roettiers del. Ravenet Sculp. A Paris chés Odieuvre M^{d} déstampes, Quai de l'Ecole vis-à- vis la Samaritaine à la belle Image. C.P.R.' [Simon François Ravenet, le vieux (1706–1774)] [**DES**; **NPG**; **WL**]

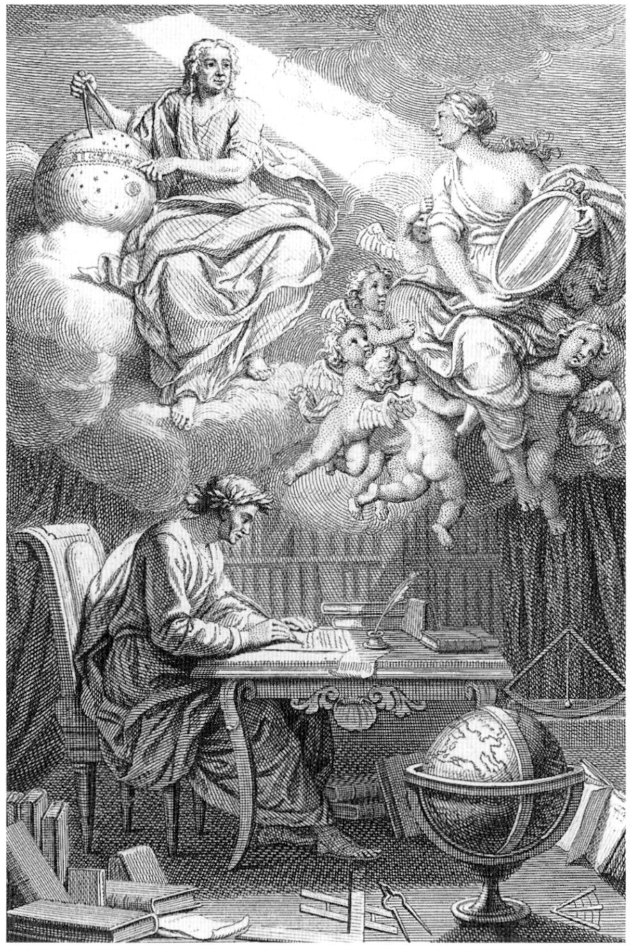

LVIII-1

LIX-2 Engraving, 20.3 × 15 cm ($8 \times 5\frac{7}{8}$ in), 'Newton. Dessiné et Gravé par Aug. S Aubin.' [Augustin de Saint-Aubin (1736–1807)] [**DES**]

LIX-3 Lithograph, 15.6 × 10.6 cm ($6\frac{1}{8} \times 4\frac{3}{16}$ in), 'Newton. Publié par Soetens & Fils, à La Haye. G.F.E.' [**DES**]

LX. Reworked plate of Bernard Picart (1673–1733), unknown date

LX-1 In 1707 Picart published an engraving entitled *La Verite Recherchée par les Philosophes* in Paris. 'B. Picart invenit et Sculpsit anõ 1707.' The engraving with a long text below is contained within an elaborate border, with copies in the Bibliothèque National and the British Museum. Francis Haskell in *Past and Present in Art and Taste*, London, 1987, reproduced alongside the Picart print an unattributed print entitled *Truth Sought by Philosophers*, of which no copy has been traced. This is from a reworking of Picart's plate, with an English text which is virtually identical with the French one except that the name of René Descartes (1596–1650) has been changed to that of Sir Isaac Newton. In the Picart engraving Philosophy, represented by a woman, leads Descartes to the knowledge of Truth, but in the English reworking there is one crucial alteration. Descartes's head has been changed to Newton's (as circled below). When the re-engraved print was published is unknown, but was almost certainly after Newton had died.

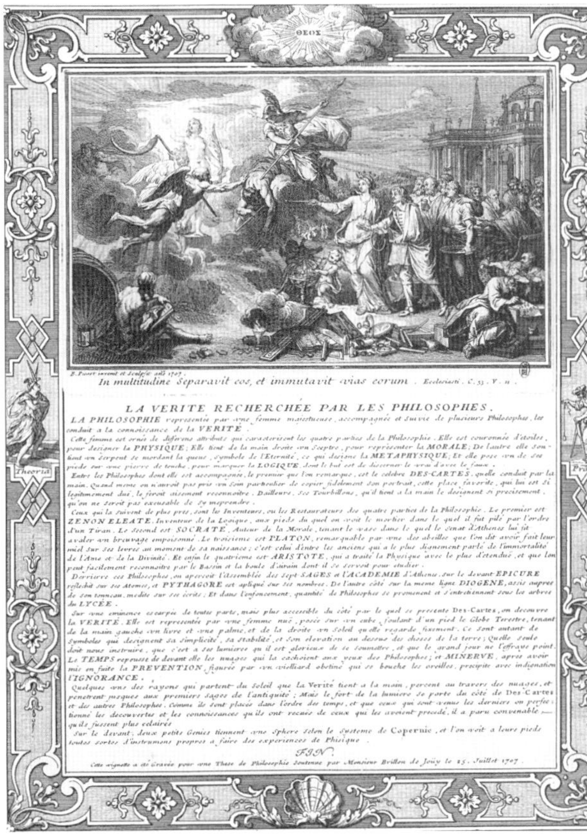

LX-1

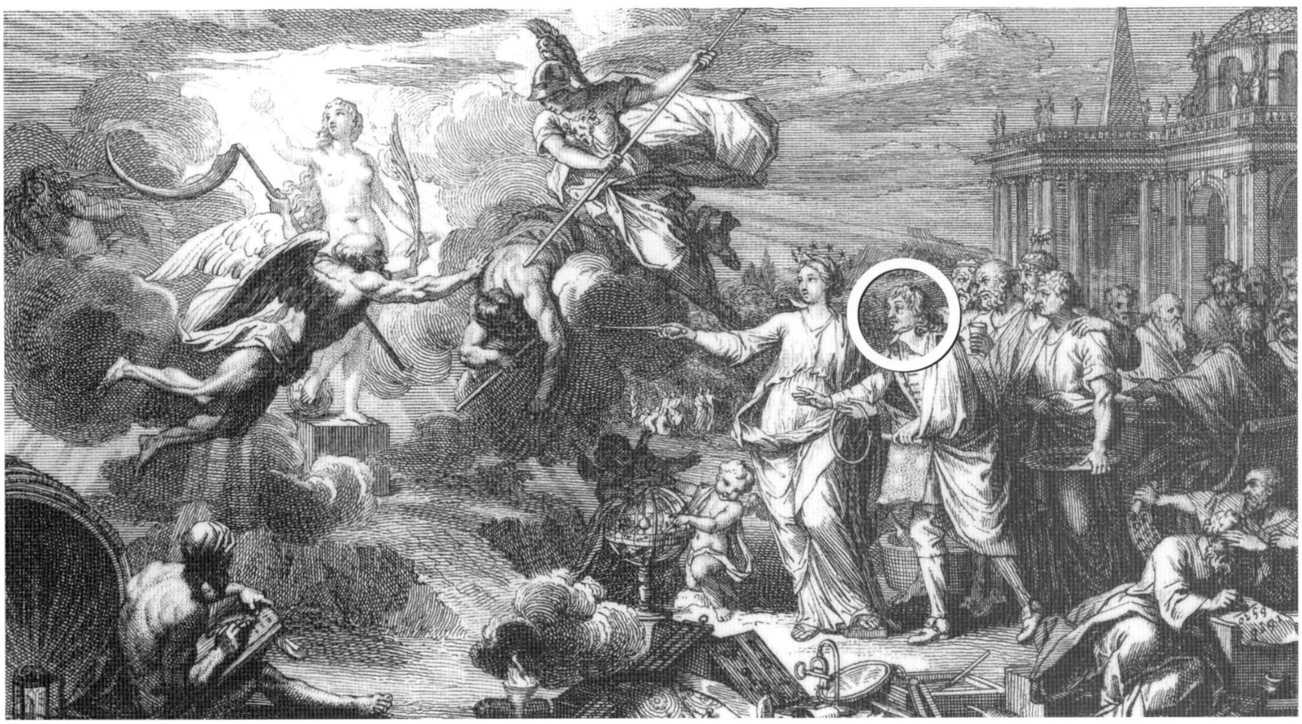

LXI. James Barry, RA (1741–1806), 1795

LXI-1 Etching and engraving, 73.2 × 47.6 cm ($28\frac{7}{8}$ × $18\frac{3}{4}$ in), after one of the six mural paintings on canvas of 'The progress of civilisation' painted by Barry from 1777 to 1783 for the Society of Arts, London. Above, Sir Francis Bacon, Copernicus, Galileo (holding telescope) and Newton (with both hands raised) behind a celestial globe with an archangel pointing across it. Below, seated or lying, Thales of Miletus, Descartes, Archimedes, Grosseteste, Bishop of Lincoln, and Roger Bacon. At foot of plate is lettering: 'This Groupe Exhibiting the Future Enjoyment of Reserved Knowledge where an Arch Angel is explaining a Solar System to the admiring Newton, Gallileo [*sic*], Copernicus & Bacon is taken from the Picture of the State of Final Retribution & with the most Respectful, Affetionate [*sic*] Regard is Dedicated to the Society for the Encouragement of Arts, Manufactures and Commerce by James Barry R.A. Professor of Painting to the Royal Academy. Paint.d Eng.d & Publ.d May. 1. 1795 by JB.'

APPENDIX

Translation from the French of 'Addition M, pp. 68–70' with plate XII from:

Essai sur la Physiognomie
destiné
A faire Connoitre l'homme et à le faire Aimer
[*Physiognomische Fragmente*, 4 vols, Leipzig, 1775–78]
par
Jean Gaspard Lavater [1741–1801]
citoyen de Zurich et Ministre du St. Évangile.

Imprimé À la Haye chez I. van Cleef, Libraire

Quatrième Partie
1803

'ISAAC NEWTON'

I have already commented how resemblance in a portrait depends on the style and talents of the painter; how the same face can be variously transformed and disfigured by an imperfect artist. The images of Newton in Plate XII verify this.

This plate is composed of engravings made from four of Newton's portraits. Each shows a great man, an extraordinary man, but each represents him in a very different light. [The four engravings, drawn and engraved by Johann Lips in 1778, are: **No. 1** (**XLIX-2**), after Seeman, 1726 (**XIV-1**), when Newton was eighty-three; **No. 2** (**XLVII-3**), after Vanderbank, 1726 (**XIII-2**), when Newton was eighty-three; **No. 3** (**XLIII-11**), after Kneller, 1702 (**III-1**), when Newton was fifty-nine; **No. 4** (**LIV-2**), after John Faber mezzotint *c.*1736 (**LIV-1**), after Guelfi bust 1732 (**A.9**).]

No. 1 In this we can recognise that the look shows an internal force, which does not just illuminate its subject, but grasps it and embraces it. It does not just lodge its gaze in our memory, but incorporates the enormous mass of ideas which the head encloses. The eyes are full of creative force, and the eyebrows bear the character of a fertile, solid and luminous spirit. I can make out from the forehead the great capacity and richness of his ideas, which strike one more by the elevation and impetuosity of the thought than by the abstraction of the reasoning. We can see an irresistible thought process produced by assurance and conviction.

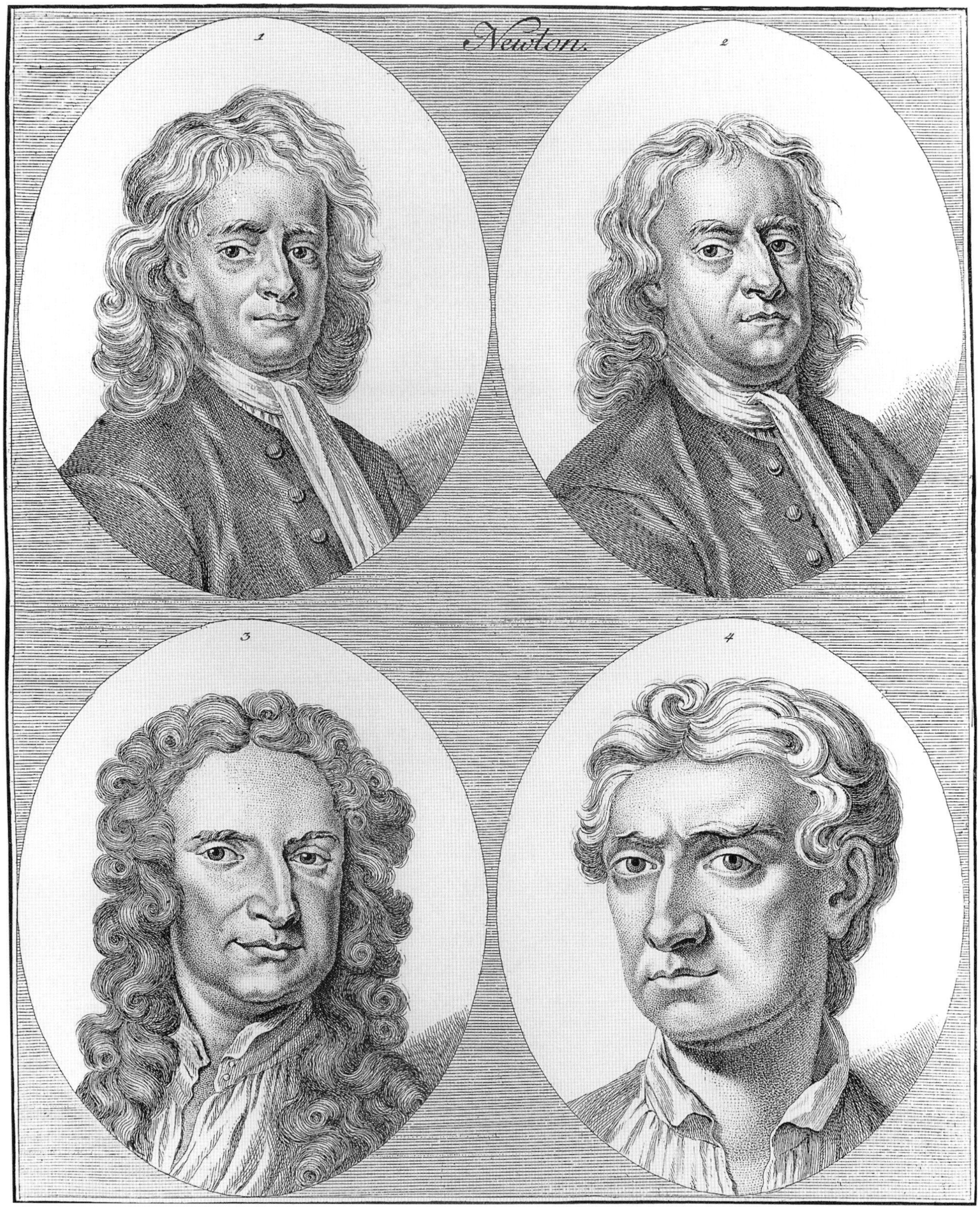

Plate XII from *Essai sur le Physiognomie* by J. G. Lavater.

Look next at the energy shown in the nose, the calm and sweetness residing in the lips without the least trace of insipidity, the solidity and probity of the chin. The right cheek is neither too relaxed nor too tense. No. 1 inspires more confidence than the other three portraits, without being so imposing as No. 2. In general, I admire the calm and purity of the whole: it is a masterstroke to have combined both of these with so much force and concentration.

No. 2 The crude effect and the vulgar appearance of this portrait come from the irregularity of the engraving. But nevertheless it is still shows an imposing physiognomy. The forehead is high and more prominent than in No. 1. The nose and mouth are more firm and severe.

You see here a republican who governs without commanding, who is accustomed to opposition, who is used to business, and who knows how to deal with and handle his affairs. A man who is firm in his enterprise and thoughtful in his plans, which he is sure of being able to carry out because he is sure of himself and the superiority of his powers. However, the nose is too fleshy and not expressive enough, and the flaring of the right alum of the nostril almost makes it look repulsive.

No. 3 The third portrait, especially from its gaze, makes one think of the professional man, or rather a man working in his study. The forehead is vaguely drawn, but more open and richer than in the previous two portraits. It implies a brain with an ability to hold a greater number of thoughts, but not one that is strongly or deeply retentive. In order to give an appearance of a man capable of deep thought, the skin below the eyes should be tauter, and the head is not sufficiently domed to suggest someone with a great memory.

Indeed, in comparing this portrait with nos. 1 and 2 we find that the eyebrows, which are placed closer to the eyes, have lost some of their energy. The nose contrasts with the openness of the forehead: it is more sanguine and weaker. The smile on the lips seems to be stimulated less from an inner happiness than from some external interest. The right cheek is unacceptable, and the appearance of the head affected and mannered, so that the whole design is tired, weak and mean.

No. 4 The fourth image tries to imitate the antique, but is treated untruthfully, though with great effect. Such a man should combine consummate knowledge with untiring energy. The forehead, however, seems compressed by the memories of past activity, while his existing faculties anticipate his future needs. What a distance from no. 3.

The expression differs between each of the four images as much as flesh differs from marble. The eyebrows are worthy of the creator of a new philosophy. The eyes assert an inner strength and speak the language of truth: their gaze is not marred by wishful thinking. The nose as a whole indicates energy, resourcefulness and wisdom. It is only the tip and the ala of the nose that show a lack of homogeneity. Finally, the lips keep under control the great strengths on which the character is based. Nos. 1 and 2 resemble each other most closely at the top, but least closely below. Nos. 3 and 4 are two extremes, one showing great relaxation in tension, the other great tension. I emphasise that the four images are only copies, but still they each show a man whose existence will live on over the centuries to come.

BIBLIOGRAPHY

1. Brian Allen. *Francis Hayman.* New Haven: Yale University Press, 1987, pp. 50–3.
2. Francis Atterbury. *The Epistolary Correspondence,* 5 vols. London, 1783–90, vol. I, pp. 180-2.
3. R. B. Beckett. *English Master Painters: Lely.* London: Routledge & Kegan Paul, 1951; C. H. Collins Baker. *Lely & the Stuart Portrait Painters.* 2 vols. London: Medici Society, 1912, vol. II, pp. 122–49; Oliver Millar. *Sir Peter Lely 1618–1680,* London: National Portrait Gallery, 1978.
4. D. Bindman. *The Catalogue of the Blake Collection in the Fitzwilliam Museum.* Cambridge, 1970, pp. 17 and 18; M. Butlin. *Catalogue of the Works of William Blake in the Tate Gallery,* London, 1957, p. 41.
5. F. E. Brasch. 'Newton's portraits and statues'. *Scripta Mathematica* 1941; 8: 199–227.
6. Sir David Brewster. *Memoirs of the Life, Writings, and Discoveries of Sir Isaac Newton.* 2 vols. 1st edn. Edinburgh: T. Constable, 1855, vol. II, pp. 396 and 414–15, note (3) quoting letter from William Stukeley to John Conduitt, 22 July 1727, Keynes MS 136, King's College Library.
7. J. T. Brighton. 'William Peckitt's Commission Book'. *Walpole Society* 1988; 54: 395–6.
8. C. H. Collins Baker, *Lely & the Stuart Portrait Painters.* 2 vols. London: Medici Society, 1912, vol. II. Appendix I: Dahl, pp. 178–81; Wilhem Nisser. *Michael Dahl and the Contemporary Swedish School of Painting.* London: John Lane, The Bodley Head and Uppsala: Almquist & Wiksells, 1927, p. 119; Catalogue of the Collection of Edward Earl of Oxford sold at auction by Mr Cock at his house in the Great Piazza, Covent-Garden on 8–13 March 1742, 'Heads of Eminent Men, no. 15: Sir Isaac Newton by Mr. Dahl', 12 March.
9. *A Descriptive Catalogue of the Grace K. Babson Collection of the Works of Sir Isaac Newton and the Material Relating to Him in the Babson Institute Library, Babson Park, Massachusetts.* With an introduction by Roger Babson Webber. Published by Herbert Reichner, New York, 1950; *A Supplement to the Catalogue of the Grace K. Babson Collection of the Works of Sir Isaac Newton and Related Material in the Babson Institute Library, Babson Park, Massachusetts.* Compiled by Henry P. Macomber. Published by Babson Institute, 1955.
10. Catalogue of exhibition: *The European Fame of Isaac Newton* at the Fitzwilliam Museum, Cambridge, 22 November 1973 – 6 January 1974.
11. *Charters, Wills etc, Trinity Colledge,* no date, pp. 208, 209, 342.
12. A. M. Crinò. 'Documents relating to some portraits in the Uffizi and to a portrait at Knole'. *Burlington Magazine* 1960; 102: 257–60; *Firenze e l'Inghilterra rapporti artistici e culturali dal XVI al XX Secolo. Saggio storico della A. M. Crinò. Catalogo critico di Mary Webster.* Florence: Gallerie Palazzo Pitti, 1971: David Paton no. 116.
13. A. W. F. Edwards. 'Is the frontispiece of *Gulliver's Travels* a likeness of Newton?' *Notes Rec. R. Soc. Lond.* 1996; 50: 191–4.
14. K. A. Esdaile. *Roubiliac's Work at Trinity College Cambridge.* Cambridge: Cambridge University Press, 1924, pp. 2–5; *The Life and Works of Louis François Roubiliac.* Oxford: Oxford University Press, 1928, p. 100.
15. Oliver Goldsmith. *The Life of Richard Nash, of Bath, Esq.* 2nd edn. London and Bath: J. Newbery and W. Frederick, 1762; A. Friedman (ed.). *Collected Works of Oliver Goldsmith.* Oxford: Clarendon Press, 1966, vol. III, p. 344.
16. J. W. Goodison. *Catalogue of Cambridge Portraits.* vol. I: *The University Collection.* Cambridge: Cambridge University Press, 1955, no. 24, pp. 21–2.
17. P. J. Grosley. *A Tour to London: or, New Observations on England and its inhabitants.* Translated from the French by Thomas Nugent in 3 volumes. Dublin, 1772, vol. I, pp. 197–8.
18. *Memoirs of Thomas Hollis Esq.* 2 vols. London: privately printed, 1780, vol. I, p. 122; vol. II, p. 800.

19. G. Keynes (ed.). *Blake's Pencil Drawings.* 2nd series, Nonesuch Press, 1956, no. 8.
20. Abraham Langford. *A Catalogue of the genuine and capital Collection of Pictures by the most celebrated MASTERS, Of that late Great and Learned Physician, Doctor Richard Mead, Deceased, will be sold by auction, at Mr Langford's, Great Piazza, Covent-Garden on 20th, 21st, and 22d of this instant March 1755.* London, 1755.
21. David Loggan. *Cantabrigia Illustrata,* 1688.
22. D. McKitterick (ed.). *The Making of the Wren Library.* Cambridge: Cambridge University Press, 1995, pp. 25–6, 130, and plates 13 to 15.
23. Oliver Millar. *The Tudor, Stuart and Early Georgian Pictures in the Collection of Her Majesty the Queen.* 2 vols. London: Phaidon Press, 1963, p. 27, nos. 332, 363, 377, 501, 642; E. Law. *A Historical Catalogue of the Pictures in the Royal Collection at Hampton Court.* London: G. Bell & Sons, 1881, no. 846.
24. Jane Munro. *John Downman 1750–1824.* Cambridge: Fitzwilliam Museum, 1996, p. 88.
25. Isaac Newton. *Philosophiae Naturalis Principia Mathematica.* 1st edn. London; *Jussu Societatis Regiae ac Typis Josephi Streater. Prostat apud plures Bibliopolas.* 1687; 2nd edn, 1713; 3rd edn, 1726, frontispiece.
26. Isaac Newton. *Opticks: or, a Treatise of the Reflexions, Refractions, Inflexions and Colours of Light with Treatises of Species and Magnitude of Curvilinear Figures.* 1st edn. London: Smith and Walford, 1704.
27. *The Correspondence of Isaac Newton.* Edited by H. W. Turnbull, J. F. Scott, A. R. Hall and L. Tilling. 7 vols. Cambridge: Cambridge University Press, 1959–77. Letter from John Collins to Newton, 5 March 1677, II, pp. 200–1; correspondence between S. Pepys and Newton, 1688–94, III, pp. 279, 281–2, 293–6, 298–301, 302–3; letter from R. Bentley to Newton, 20 October 1709, V, p. 8; letter from Roger Cotes, 20 July 1712, V, pp. 315–16; P. Varignon, quoted in VII, p. 106, note (2).
28. J. D. Passavant. *Tour of a German artist in England: with notices of private galleries, and remarks on the state of art by M. Passavant.* 2 vols. London: Saunders and Otley, 1836.
29. David Piper. *Catalogue of 17th Century Portraits in the National Portrait Gallery, 1625–1714.* Cambridge: Cambridge University Press, 1963, pp. 248–51.
30. Robert Robson. 'Cipriani Window'. *Trinity Review,* 1974: 14.
31. J. D. Stewart. *Sir Godfrey Kneller and the English Baroque Portrait.* Oxford: Clarendon Press, 1983, p. 120, nos. 528, 529, 530; C. H. Collins Baker. *Catalogue of the Petworth Collection of Pictures in the Possession of Lord Leconfield.* London: Medici Society, 1920, p. 66, no. 153.
32. William Stukeley. *Memoirs of Sir Isaac Newton's Life.* Edited by A. Hastings White. London: Taylor and Francis, 1936, pp. 12–13, 64, 84–6, in text and in footnote.
33. George Vertue. *The Notebooks of George Vertue* I–VI. *The Walpole Society,* vols. 18, 20, 22, 24, 26, 30; index of I–V in 29, 30. Oxford, 1930–55. Vertue *Notebook* I: 18, p. 138; IV: 24, pp. 5, 56, 146.
34. F. M. Arouet de Voltaire. *Elémens de la Philosophie DE NEUTON mis à la portée de tout le monde par M^R. DE VOLTAIRE, à AMSTERDAM chez Etienne Ledet & Companie,* 1738, frontispiece; R. L. Walters. 'The allegorical engravings in the Ledet-Desbordes edition of the *Eléments de la philosophie de Newton.*' in R. J. Howells, A. Mason, H. T. Mason, D. Williams (eds.), *Voltaire and His World.* Oxford: Voltaire Foundation, Taylor Institution, 1985, pp. 27–49.
35. G. F. Waagen. *Works of Art and Artists in England.* Trans. H. E. Lloyd. 3 vols. London, 1838; *Treasures of Art in Great Britain being an account of the Chief Collections of Paintings, Drawings, Sculptures., Illuminated MSS etc, etc, by Dr Waagen.* Trans. Lady Eastlake. 3 vols. London: John Murray, 1854–7, III, pp. 7–17.
36. Horace Walpole. *Anecdotes of Painting in England: with some account of the principle artists; and incidental notes on other arts; Collected by the late Mr George Vertue And now digested and published from his original MSS. By Mr Horace Walpole.* 4 vols. Printed by Thomas Farmer at Strawberry-Hill. 1762–71.
37. R. S. Westfall. *Never at Rest: A Biography of Isaac Newton.* Cambridge: Cambridge University Press, 1980, pp. 180–3, 481, 631, caption fig. 13.1;

M. Beretta. 'At the source of western science: the organization of experimentalism at the Accademia del Cimento (1657–67)'. *Notes Rec. R. Soc. Lond.* 2000; 54: 131–51; M. Keynes. 'A bogus Newtonian "curiosity" at the Royal Society'. *Notes Rec. R. Soc. Lond.* 2001; 55: 247–52.

38. R. E. W. Maddison. 'The portraiture of the Honourable Robert Boyle, F.R.S.' *Annals of Science* 1959; 15: 141–214 with 55 plates, p. 157; Edward Wright. *Some Observations made in Travelling through France, Italy, &c., In the Years 1720, 1721 and 1722.* 2 vols. London, 1730; *Gli Uffizi Catalogo Generale.* 2nd edn. Florence: Centro Di, 1980; H. Acton. *The Last Medici.* London: Faber and Faber, 1932, pp. 109–11; L. Magalotti. *Un principe di Toscana in Inghilterra e in Irlanda nel 1669. Relazione ufficiale del viaggio di Cosimo de' Medici tratta dal 'Giornale' di L. Magalotti, con gli acquerelli palatini.* A cura di A. M. Crinò. Rome: Edizioni di Storia e Letteratura, 1968. This account of the Prince of Tuscany's tour through England in 1669 was translated in 1821: *The Travels of Cosmo the Third, Grand Duke of Tuscany, through England during the Reign of King Charles the Second., with Memoir of his Life.* Trans. by J. Mawman. Ed. Robert Stewart. London, 1821, p. 229; W. E. Knowles Middleton (ed. and trans.). *Lorenzo Magalotti at the Court of Charles II.* Ontario: Waterloo, 1980; S. Pepys. *The Diary of Samuel Pepys, 1660–1669: A New and Complete Transcription.* 11 vols. edited by R. Latham and W. Matthews with four contributing editors. London: G. Bell & Sons, 1970-83, vol. IX, *1668–69,* 1976, 5 April 1669, p. 509; A. Cook. *Edmond Halley: Charting the Heavens and the Seas.* Oxford: Clarendon Press, 1998, p. 161–3; Royal Society Library: Letter Book Copy, 11 (1), 114, Letter from W. Rooke to E. Halley, 2 February 1688.

39. W. K. Wimsatt. *Portraits of Alexander Pope.* New Haven: Yale University Press, 1965, pp. 237–40; J. Riely and W. K. Wimsatt. 'Supplement to the portraits of Alexander Pope' in *Evidence in Literary Scholarship: Essays in Memory of James Marshall Osborn.* R. Wellek and A. Ribeiro (eds.). Oxford: Clarendon Press, 1979, pp. 150–1; E. G. D'Oench. *The Conversation Piece: Arthur Devis and His Contemporaries.* New Haven: Yale Center for British Art, 1980. Catalogue of exhibition at the Yale Center for British Art, New Haven, Connecticut, 1 October to 30 November, 1980; *The Polite Society by Arthur Devis, 1712–1787: Portraits of the English Country Gentleman and His Family.* Catalogue of exhibition at Harris Museum & Art Gallery, Preston, 1 October – 12 November 1983; National Portrait Gallery, 25 November 1983 – 29 January 1984.

40. F. O'Donoghue. *Catalogue of Engraved British Portraits Preserved in the Department of Prints and Drawings in the British Museum.* London: British Museum, 1912, vol. III, pp. 329–31.

41. D. E. Smith. 'Portraits of Sir Isaac Newton' in *Isaac Newton 1642–1727: A Memorial Volume.* Edited by W. J. Greenstreet. London: G. Bell and Sons, 1927, pp. 170–8. The David Eugene Smith Collection of Engravings of Newton is in the Rare Book and Manuscript Library, Columbia University.

42. R. Burgess. *Portraits of Doctors & Scientists in the Wellcome Institute: a catalogue.* London: Wellcome Institute, 1973.

43. Thomas Birch. *The HEADS of Illustrious persons of Great Britain engraved by Mr Houbraken, & Mr. Vertue with their LIVES and CHARACTERS by Thomas Birch, A.M. FRS* London: John & Paul Knapton, 1743, pp. 147–8, plate opposite p. 147.

44. Johannes Castillioneus. *Isaaci Newtoni / Equitis Aurati / OPUSCULA / Mathematica, Philologica / collegit partimque Latinè vertit ac recensuit / Joh. Castillioneus.* 3 vols. Lausanne & Geneva apud Marcum-Michael Bousquet & Socios, 1744, title page engraving.

45. L. Forrer. *Biographical Dictionary of Medallists.* London: Spink & Son, 1912, vol. V, pp. 159–61.

46. S. Taylor. 'Queen Caroline and the Church of England' in S. Taylor, R. Connors and C. Jones (eds.), *Hanoverian Britain and Empire.* Woodbridge: The Boydell Press, 1998, pp. 82–101; D. Bertoloni Meli. 'Caroline, Leibnitz and Clarke'. *J History of Ideas* 1999; 60: 469–86.

PART II

SCULPTURES

AD VIVUM SCULPTURES

POSTHUMOUS SCULPTURES

AD VIVUM SCULPTURES

A. BUSTS

David le Marchand (1674–1726)

A.1 Ivory bust on cylindrical marble stand, 1714. (Avery,[1] no. 47)
18.6 × 13.3 × 8.8 cm ($7\frac{3}{8} \times 5\frac{1}{4} \times 3\frac{1}{2}$ in)
National Gallery of Victoria, Melbourne, Australia (no. 4118.3).

The bust has Newton unwigged, with a bare neck, wearing a loose garment *all'antica.*

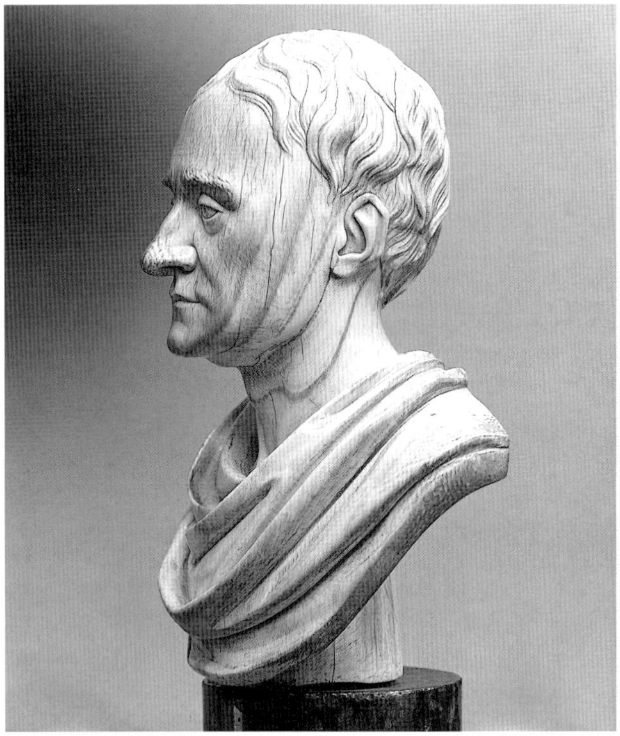

A.1

Inscribed: D. L. M.*Sct. 1714*

Provenance: Charles Montagu, Earl of Halifax, FRS (1679–1715) (sold Cocks, Covent Garden, 1739); 2nd Earl of Oxford (1689–1741), Wimpole Hall, Cambs; ? James West, PRS (1704–1772) (sale, Langford & Son, London, 6 March 1773, p. 22, lot 19); Countess of Gosford, 1930s; Alfred Spero, London; Howard Spensley, Bedfordshire, who bequeathed it to the National Art Collections Fund to be presented to the Government of Victoria, in 1939.

George Vertue[28] described this bust among the 'Pictures sold in Lord Halifax's sale': 'dito, a small bust in Ivory done from the life Sir Isaac Newton a head without a wigg – 1714 by Marchand'. Possibly commissioned by Lord Halifax.[1]

A.2 Ivory bust, 1718. (Avery,[1] no. 68)
24.8 cm ($9\frac{3}{4}$ in) high, max. width 15.1 cm (6 in)
British Museum MLA 1765, 6–29.[8]

Unwigged bust, with central parting to the hair, wearing an unbuttoned shirt under a loose garment, with particular folds over the left side of the chest.

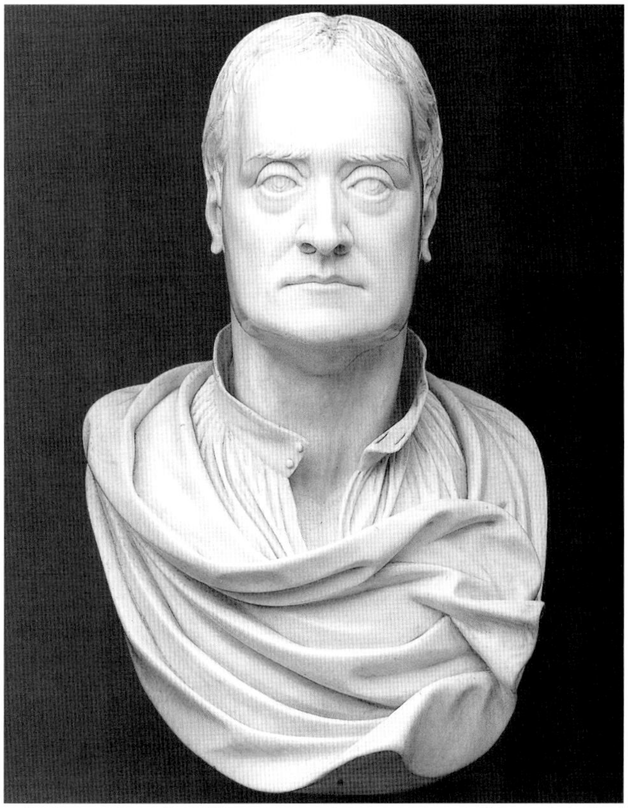

A.2

Inscribed: (on back of stand)

ISAACUS NEWTON / EQ : AVRA /
An : 1718 / le Marchand / Sc. ad Vi.

Provenance: Mathew Raper II (1665–1748), Director of the Bank of England, who may have acquired it from Newton's half-nephew, Benjamin Smith (1700–1776); Mathew Raper III, FRS (1704–1778), who gave it to British Museum in 1765.

This bust, described by William Stukeley, FRS (1687–1765) in his *Memoirs of Sir Isaac Newton's Life*[36] as 'a very good likeness', which he later saw after Newton's death 'in possession of one of his heirs, Mr. Ben Smith', was carved from life in 1718, using ivory taken from the broader lower zone of an exceptionally large elephant tusk. The carving picks out Newton's square jaw, blunt nose and squarish, balding cranium, and was commissioned at a cost of one hundred guineas by Newton, as stated by Benjamin Smith when:

> left a small ivory bust of admirable workmanship by that celebrated artist Marchand, which from its elegance, similitude and placid expression is truly valuable. It is said to have cost Sir Isaac one hundred guineas and is specified in an authentic inventory of his effects, taken by virtue of a commission of appraisement in April 1727, now in my possession. (Quoted from *Correspondence of Sir Isaac Newton and Professor Cotes*[36] [Roger Cotes, FRS (1682–1716)]).

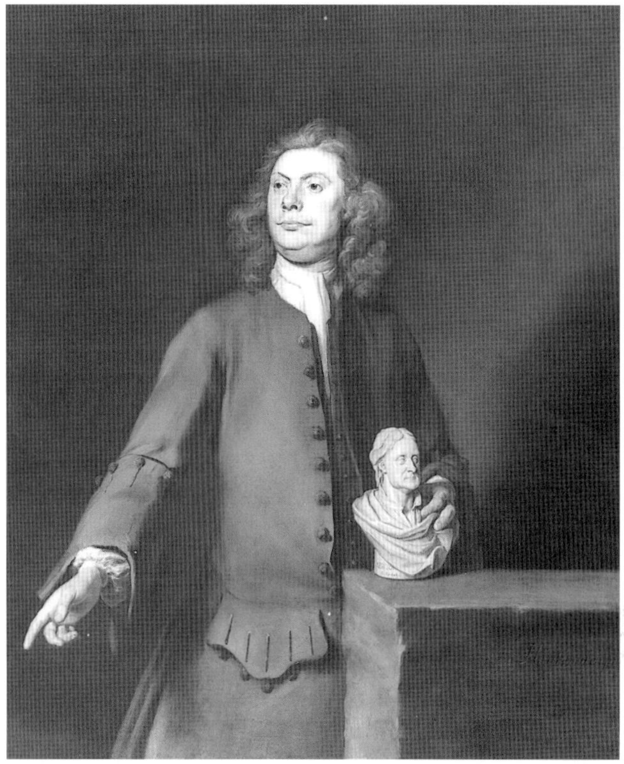

A.3

A.3 The ivory bust (**A.2**) of 1718 in the British Museum is illustrated in a portrait of *David le Marchand*, oil on canvas, 125.7 × 101.6 cm ($49\frac{1}{2}$ × 40 in), by **Joseph Highmore (1692–1780)**, painted *c.*1723. National Portrait Gallery, London (no. 6142).

A.4 Ivory bust, perhaps by le Marchand, 1720–5.
(Avery,[1] no. 69b)
19 cm ($7\frac{1}{2}$ in) high
Collection of Lord Thomson of Fleet donated to The Art Gallery of Ontario, Toronto, in 2002.

Provenance: Mrs Kavan; Sotheby's, 26 March 1965 (lot 7).

B. RELIEFS

David le Marchand (1674–1726)

B.1 Rectangular ivory plaque; full-frontal in high relief, wigged, 1702–10. (Avery,[1] no. 44)
19.75 × 13 cm ($7\frac{3}{4} \times 5\frac{1}{8}$ in)
Lord Thomson of Fleet, Toronto.

Inscribed:

D. Le Marcha[d] adViv⁻ Sc[lt].

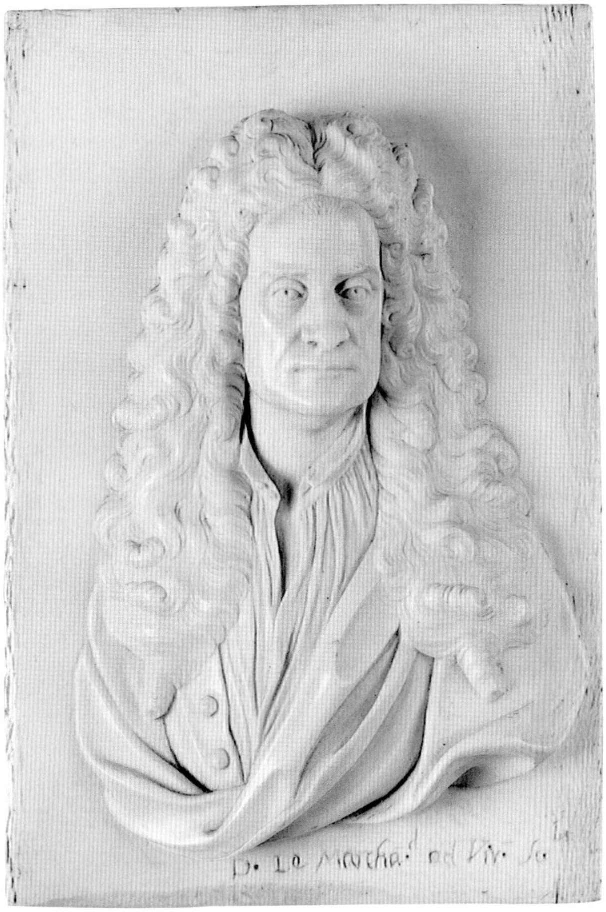

B.1
see colour Plate XIII

Provenance: Dr Richard Mead, MD, FRS (1673–1754) who was Newton's physician; sold 11 March 1755 by Abraham Langford (1711–74) in association with Samuel Baker in *The Museum Meadianum Pars Prior Sale,* when described as 'a *D. L. Marchand* ad vivum sculpta'; Edward Poore's sale at Richardson's, 31, The Strand, London, 19 April 1805 (lot 49); Sotheby's, 20 December 1963 (lot 165).

Described by George Vertue[28] as 'an Alto-relieve with the wigg on ... by Marchand this last I think the best & in poses Dr. Meade'.

B.2 Wax, full-frontal, high-relief portrait, 1702–10. (Avery,[1] no. 45)
15.2 cm (6 in) high
Trustees of the Corsham Estate, Corsham Court, Wiltshire.

Provenance: Francis Hall (? son of Thomas Hall (d.1718), Chief Clerk to Newton at the Mint) of Sandford House, Crediton, Devon; Sir John Chichester (who married widow of Francis Hall in 1733), to his nephew, Revd John Sanford, and then by descent to Lord Methuen.

This is either the model for, or a cast of, **B.1**.[1]

B.3 Ivory medallion, three-quarter profile to the left, wearing a wig, 1702–10. (Avery,[1] no. 46)
Oval, 24.2 × 20.4 cm ($9\frac{1}{2} \times 8$ in)
Lost.

Provenance: J. W. Brett sale, Christie's, 5 April 1864 (lot 2035), described as 'An alto-relievo bust of Sir Isaac Newton, by Denis [*sic*] le Marchand, in a black frame-oval', with a photograph.

C. NEWTON'S DEATH MASK

Katharine Esdaile (1881–1950)[9,11] claimed that the original cast of the death mask was taken under the instructions of John Conduitt by Michael Rysbrack (1694–1770), who was then commissioned to carve the monument in Westminster Abbey (**D.1–D.6**), unveiled in April 1731. Rysbrack, born in Antwerp, arrived in London in 1720, but it is not known whether he ever met Newton, who died 20 March 1727.

Esdaile based her entire claim on a section: 'On Sculpture' in a book[11] *État des Arts en Angleterre* (1755) by Jean André Rouquet (1701–58), a painter in enamel and friend of Hogarth, but this is hardly confirmed on reading what Rouquet wrote. He mentions Rysbrack had the honour *of being employed in decorating the tomb of Sir Isaac Newton*, and (in emphasising that it was not erected by the public or the nation) that *Mr Conduit, who married that Great Man's niece, was at the whole expence of the monument. It must be allowed that sculpture could be hitherto considered in England only as a funeral talent, having scarce ever been used but for the decoration of tombs. It is only within these few years that it has been more generally applied.* The latter sentence refers to the carving and displaying of portrait busts independently of church monuments.

We do not know who took the death mask, and it is significant that George Vertue (*Notebooks* III: p. 57[28]), in discussing Rysbrack's bust of Newton, wrote: *1732. Mr. Michael Rysbrake did Sr. Isaac Newton immediately after his death from pictures or draughts* without mentioning the death mask. It can only be surmised whether Rysbrack used the death mask for carving the Conduitt bust (**A.10**) and the bust at Stowe (**A.12**) and the figure (**D.4** and **D.5**) in the monument in Westminster Abbey. But when L. F. Roubiliac (?1705–1762), born in Lyons and who only arrived in London in 1730, made the terracotta model (**A.18**) for his bust of Newton (**A.19**) and later the statue (**D.7**) in Trinity Chapel, he did use a copy (**C.2**) of the death mask.[9,10,11]

The taking of a death mask is done when the body is still warm. Plaster of the consistency of soup is ladled over the face to cover it a few millimetres. thick. A thread is then placed over the middle of the forehead, the bridge of the nose, the mouth and chin. More solid plaster is spread over the first layer to provide a firm outer shell, and before it sets the thread is pulled forward, dividing it into two halves. As soon as the outer layer has set hard, the halved moulds are detached from the head and immediately fitted together. The inside of the piece mould is cleaned and smoothed, and then filled with plaster to obtain the finished death mask on smoothing of its surface.[11] The join down the centre of the mask will often be obvious.

C.1 White plaster cast, with join down the centre of the face, varnished at some time, damaged, 1727.
Cast 19.5 × 14.4 cm ($7\frac{3}{4} \times 5\frac{5}{8}$ in)
Keynes Library, King's College, Cambridge.

By descent from John Conduitt in the collection of the Earls of Portsmouth; purchased by J. M. Keynes (1883–1946) at the Viscount Lymington Sale of The Newton Papers, Sotheby's, 14 July 1936 (lot 332); bequeathed to King's College in 1946.

Exhibited: *Footprints of the Lion: Isaac Newton at Work*, Cambridge University Library, 9 October 2001 – 23 March 2002.

C.2 White plaster cast with join down centre of face.
20 × 14 cm ($7\frac{7}{8} \times 5\frac{1}{2}$ in)
Royal Society.

This cast belonged to Roubiliac, and was sold at the auction sale of his studio contents at his 'dwelling-House in St Martin's Lane' by Mr Langford on 12–15 May 1762 to a dealer in statues in Tichborne Street; there it lay disregarded for half a century or more until discovered and purchased by Samuel Hunter Christie, FRS (1784–1865), who presented it to the Royal Society in 1839.

C.3 White plaster cast with join down centre of the face.
19.5 × 14.4 cm ($7\frac{1}{2} \times 5\frac{5}{8}$ in)
Wren Library, Trinity College, Cambridge.

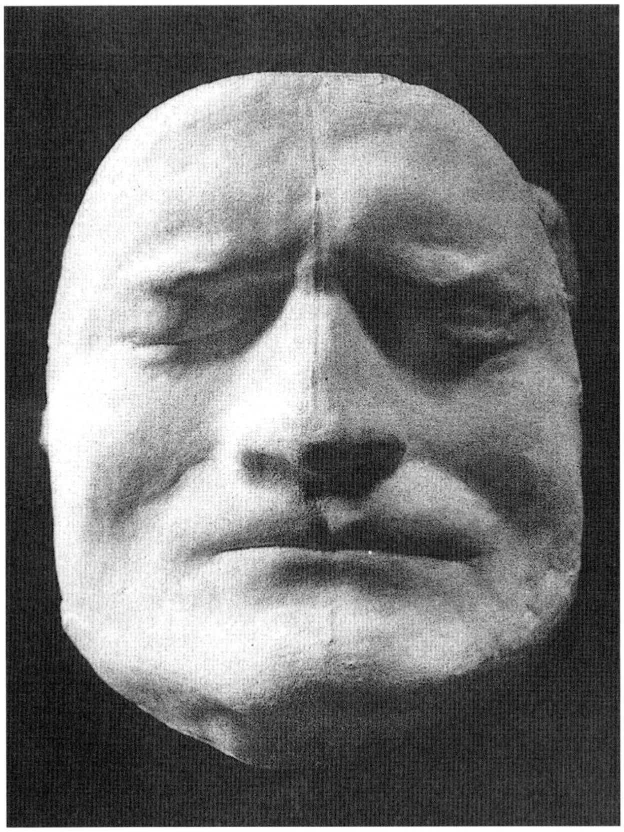

C.3

This cast, that also belonged to Roubiliac, is kept in a box which bears an old label: 'Given by Roubiliac to his pupil Nathan Smith. Mask taken from Sir Isaac Newton immediately after his death. Used by Roubiliac in making the statue [**D.7**].' Nathaniel Smith (*c.*1741– after 1800) was apprenticed to Roubiliac in 1755. It seems more likely, from note by M. Cormack of 25 January 1974 in Trinity Library, that it is **C.2** that has the better claim to have been the one used by Roubiliac.

C.4 White plaster cast with join down centre of the face.
21.5 × 12.7 cm ($8\frac{1}{2} \times 5$ in)
Hunterian Museum and Art Gallery, Glasgow (no. C.44).

Presented to the Museum in 1819 by John Flaxman (1755–1826).

C.5 Plaster cast on circular base, lower half painted black.
24.8 × 14.7 cm wide ($9\frac{3}{4} \times 5\frac{3}{4}$ in)
Grace K. Babson collection (no. 519.686) on permanent loan to the Burndy Library, Dibner Institute, Massachusetts Institute of Technology.

Given in 1834 by Mrs Joseph Coolidge, Jr, granddaughter of Thomas Jefferson (1743–1826), to whom it had almost certainly belonged, to the Boston Athenaeum, whose Trustees presented it to the Babson Institute, Wellesley, Massachusetts in 1951.

C.6 Plaster cast: collection of Adam Henss (1780–1856); Archenhold Observatory, Treptow, Berlin. Benkard[11] reports another copy of the death mask in Germany in the Collection of Death Masks made by Karl Gustav Carus (1789–1869) and bequeathed to the Philosophical Faculty of Leipzig University, which presented it to the University Anatomical Institute.

C.7 Painted plaster cast, with central join, in Collection of Death Masks[11] made by Laurance Hutton (1843–1904) mainly in the 1860s, and bequeathed to Princeton, University Library. Attached to it is Hutton's notation that it was 'From the original mask by Roubilliac [*sic*] in the rooms of the Royal Society.'

POSTHUMOUS SCULPTURES

A. BUSTS

After David le Marchand (1674–1726)

A.5 Ivory bust, by a copyist after le Marchand (**A.4**), 1725–50. (Avery,[1] no. 69d)
20 cm ($7\frac{7}{8}$ in) high
Collection of Lord Thomson of Fleet donated to The Art Gallery of Ontario, Toronto, in 2002.

Inscribed (on the support at the back):
ISAA. NEWTO. Eq. A.G.

Provenance: Purchased from Alfred Speelman in 1953.

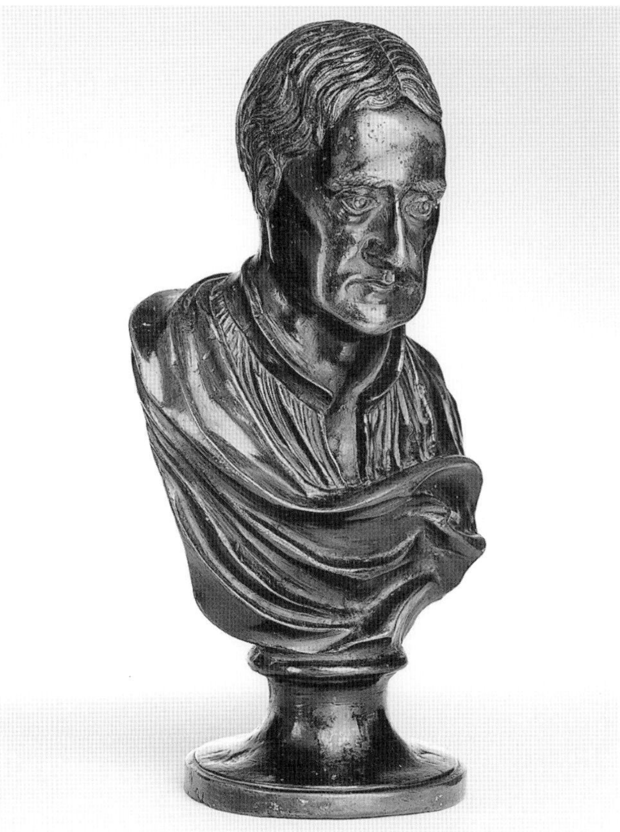

A.6

A.6 Bronze bust on round socle, after **A.2**, early eighteenth century.
24.8 cm ($9\frac{3}{4}$ in) high
Roy Davids, Great Haseley, Oxfordshire, 1997.

Giovanni Baptista Guelfi (1691– after 1734)

A.7 Model in terracotta of bust of Newton, 1731, now missing.[14]

In 1731, Queen Caroline (1683–1737) ordered four busts in stone for her Grotto, also called The Hermitage, 'a grotesque building of a heap of Stones laid rudely together, which seems as if it had stood many hundred years, though but lately done', designed by William Kent (1684–1748), the architect, and built for her in 1730 at Richmond; demolished in 1775. George Vertue (1684–1756) wrote in his *Notebooks* III: p. 51[28] in 1731: 'Four busts of Stone are to be made by Sign^{r} Guelphi, a Sculptor for the Queens … at Richmond. S^{r}. I. Newton Lock D^{r}. Clarke & Mr Woolaston – this Sig. Gulpha workes under the direction of Mr. Kent at the House of Lord Burlington.' Guelfi, who came from Rome to work for Lord Burlington (1695–1753) in 1714, made models (presumably in terracotta) for the four busts: there is a record that he received a series of payments from Richmond Old Lodge in 1731 and 1732.[14] When he left England for Bologna in 1734, he gave the four models to William Kent, who bequeathed them to Lady Isabella Finch, but since then they have been missing.

In 1733 Francis Walsingham wrote in the *Free Briton* (no. 195, 16 August 1733) in praise of Queen Caroline as a patron of the arts and especially sculpture, and how she showed 'her *peculiar Affection* to this Country … Whilst Bacon and Boyle, Sir Isaac Newton and Dr. Clarke, Locke and Wollaston, employ the Hand of Rysbrack, and are placed in *her Majesty's* Grotto, not even her own [German] *Leibnitz* [Gottfried Wilhelm Leibniz, FRS (1646–1716)] is allowed a Place there.' Three weeks later, *The Grubstreet Journal* (no. 193, 6 September 1733) pointed out that Walsingham had made 'several historical mistakes and one egregious blunder which overturns his whole panegyric, and entirely destroys the reputation in the art of statuary. For, in order to do honour to Mr Rysbrack, he has attributed to him the [marble] bustoes in *her Majesty's* Grotto; which unfortunately happen to be the Work of another, and as Some think, a much inferior Hand [that of Guelfi]'.

George Vertue (*Notebooks*, III, p. 66)[28] quotes the 1733 *Free Briton* passage listing six busts by Rysbrack in the Queen's Grotto, but unfortunately without including the correction in *The Grubstreet Journal*. This has led to erroneous presumptions and attributions, which can now be corrected.

Five of the six named marble busts, now housed in Kensington Palace, show a similarity in style of both dress and physiognomy that is unlike that of Rysbrack, but can be attributed to Guelfi. The bust of Newton (**A.9**) is nearly identical to the stone bust (**A.8**) originally in the possession of Alexander Pope and named by him in his will as by Guelfi. The bust of Dr Samuel Clarke (1675–1729) was confirmed as carved by Guelfi in the postscript to a letter from Michael Rysbrack to Sir Edward Littleton (*c.*1726–1812), dated 20 January 1756: 'Sir I did not make the Bust of Dr. Clark in the Hermitage: It was Done By Mr. Ghuelphi an Italian, who is Dead.'[31] The face in the bust of John Locke, FRS (1632–1704) differs from that in the bust of him in stone by Rysbrack *c.*1726–9 at Stowe.[12] There is no known bust of William Wollaston (1660–1724) by Rysbrack. From the *Gentleman's Magazine* (August 1732, 2: 922), these four busts by Guelfi were placed in position in the Grotto before August 1732.

In 1732, a payment was made to Guelfi of £68 'for the Busto of the honble Robert Boyle [FRS (1627–1691)] in statuary marble with a neck of veined marble & for carriage & repairing the antique statue of Venus' – no model of this bust of Boyle is known.[14] This was the fifth bust by Guelfi, and the last one to be placed in the Grotto, which was sometime before April 1733 (*Gentleman's Magazine* 3: 207).

Giometti[14] lists the various payments to Guelfi made by the Paymaster of Richmond Old Lodge between July 1731 and August 1732 when the last payment of £68 for the fifth bust was made, but Colvin[14] found that these payments did not cover the cost of carving the first four busts. The sixth, the bust of Francis Bacon (1561–1626), was probably never carved, but if it was executed it was not placed in the Grotto;[14] a bust of Bacon recently moved from Windsor Castle to Kensington Palace is not in the same hand as the other five, nor by Rysbrack, and is probably nineteenth century.

A.8 Stone bust without socle, *c.*1731.
60.8 × 43 cm (24 × 17 in)
By descent in collection of Earl of Mansfield, Scone Palace, Perthshire.

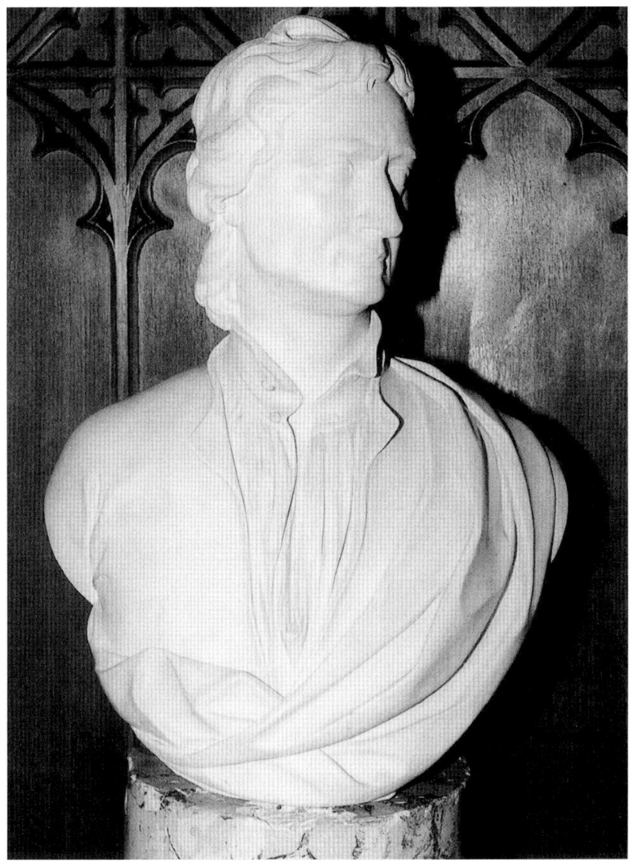

A.8
see colour Plate XIII

Newton is looking to his left, wearing a buttonless shirt under a coat, with drape over left shoulder.

It is not known how this stone bust came into the possession of Alexander Pope (1688–1744), presumably for his grotto at Twickenham, but it could have been commissioned by him (see Francis Haskell. 'The apotheosis of Newton in art', 1967[6]). In his will, Pope left Councillor William Murray (1705–1793), created 1st Earl of Mansfield in 1776, 'the head of Sir Isaac Newton by Guelfi', which was identified at Scone Palace in 1951–2[31] – the treatment of its back, rough hewn and undefined at the torso, is Italianate and not Flemish (K. Eustace, 2002). Guelfi does not appear to have made stone busts of John Locke, Samuel Clarke and William Wollaston from the other models that he gave to William Kent in 1734 – see **A.7**.

A.9 Marble bust, 1732.
60 cm ($23\frac{5}{8}$ in) high including shaped, round socle
Kensington Palace, London.

Newton looks to his left, with buttonless, open-neck shirt and coat.

Made for Queen Caroline's Grotto at Richmond, and in place there by August 1732 – see **A.7** – the marble bust is nearly identical to the stone bust **A.8**. Moved to Buckingham Palace; moved to George IV's new Grand Corridor, Windsor Castle, in 1828; Kensington Palace since 1923. The bust was engraved in mezzotint *c.*1736 by J. Faber, Jr (**LIV-1**).

As a result of its mention by Vertue[28] in 1733, it has long been wrongly claimed that it was made by Rysbrack, who, it was presumed, must have taken over a revised commission from Guelfi but still used his model. This cannot be accepted, particularly as it would have been after Rysbrack had already completed the Conduitt bust of Newton (**A.10**) *c.*1727–9, the stone bust at Stowe (**A.12**) *c.*1727-9, and the monument to Newton (see **D.4** and **D.5**) in Westminster Abbey, which was unveiled in April 1731.

A.9 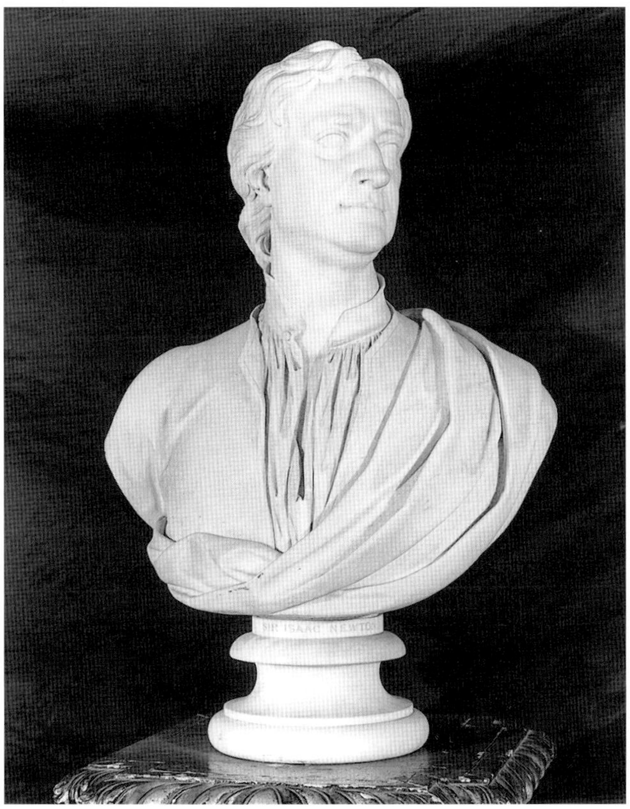

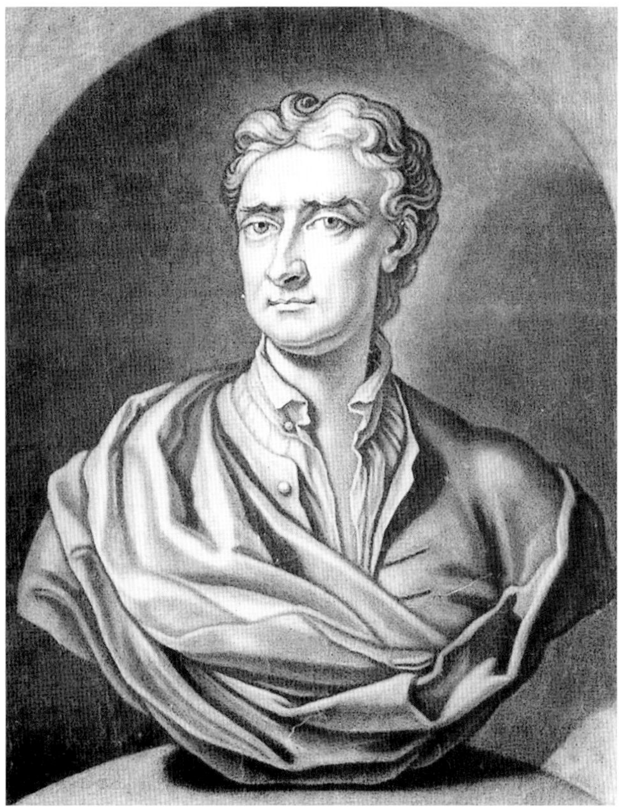 LIV–1

(Johannes Michel) Michael Rysbrack (1694–1770)

A.10 Marble bust, known as the 'Conduitt bust', with square, carved socle, *c.*1727–9.
60 × 56 × 24 cm ($23\frac{1}{2} \times 22 \times 9\frac{1}{2}$ in), socle 12.5 cm (5 in) high
Earl of Portsmouth, Farleigh Wallop, Basingstoke, Hampshire.

Head turned half-left, wearing an unbuttoned shirt and coat beneath a loose drape, mainly over right shoulder.

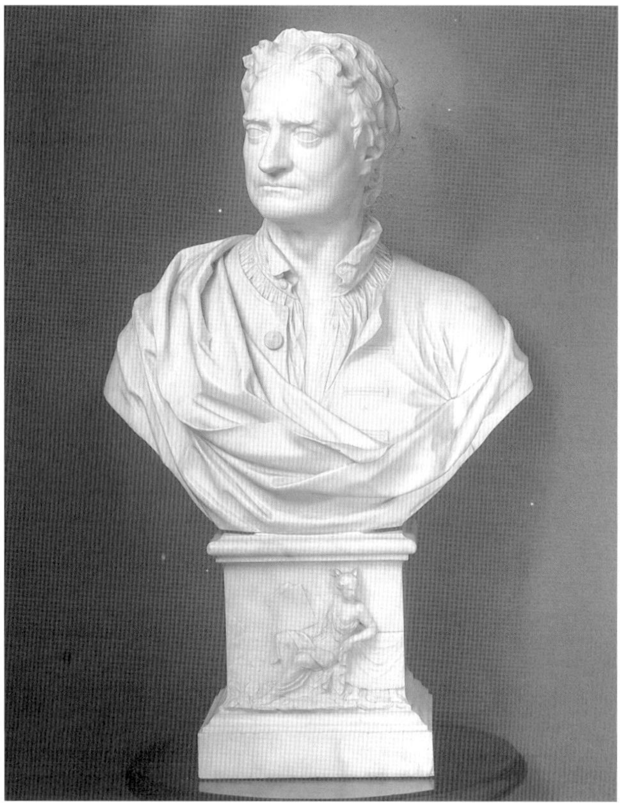

A.10
see colour Plate XIV

The socle is carved with a seated female figure with bared left breast, who cannot be identified. Malcolm Baker has unconvincingly suggested that she is related to the draped female figure, who reclines over a globe, at the top of Rysbrack's monument in Westminster Abbey (**D.5**), and even less to the muse Urania of Rysbrack's earlier, undated original design (see **D.3**).

According to George Vertue (*Notebooks*, III, p. 57[28]): '1732. M^r^. Michael Rysbrake did S^r^. Isaac Newton immediately after his death from pictures or draughts', but notably Vertue does not mention the death mask. The bust, commissioned by John Conduitt, FRS (1688–1737) and by descent to the Earl of Portsmouth, is less idealised than Roubiliac's portrait bust (**A.19**) of *c.*1732, and is in contemporary dress whereas Roubiliac's bust is not. In making **A.10** Rysbrack appears to have used the Kneller 1702 portrait (**III-1**) and the David le Marchand ivory bust (**A.2**).

The bust, for which the terracotta, or model, is missing, has been dated *c.*1731–2[6] seemingly because it appears in the scene of 1732 depicted in Hogarth's painting (**A.11**) of 1732–5. In view of Vertue's remark, an earlier date *c.*1727–9 is more likely: **A.12** in stone was also carved *c.*1727–9,[3,12] and the Rysbrack monument (**D.4**, **D.5**) in Westminster Abbey with Newton's figure was unveiled in April 1731. Katharine Esdaile claimed that the Conduitt bust was by Roubiliac, unaware of the painted plaster bust (**A.14**), signed *M. Rysbrack 1739*, which had been in Trinity College Library since 1932.[31]

A.11 A bust of Newton is illustrated in *A scene from 'The Indian Emperour' or The Conquest of Mexico by the Spaniards*, oil on canvas, 130.8 × 146.6 cm ($51\frac{1}{2} \times 57\frac{11}{16}$ in), by **William Hogarth (1697–1764)**, painted 1732–5. A detail is shown here but the full painting is shown in colour **Plate XIV**.

This painting in the collection of the family of the 6th Earl of Ilchester (1874–1959) was commissioned by John Conduitt, and shows four children, including Kitty Conduitt (1718–1750), acting in the play, written in 1665 by John Dryden, FRS (1631–1700), in Mr Conduitt's house on 27 April 1732. The play is being performed in front of the three youngest children of George II and Queen Caroline, William, Duke of Cumberland (1721–1765), and the princesses Mary (1723–1772) and Louisa (1724–1751). The audience also includes the Dukes of Montagu and Richmond, the Earl of Pomfret and Thomas Hill, the Secretary of the Board

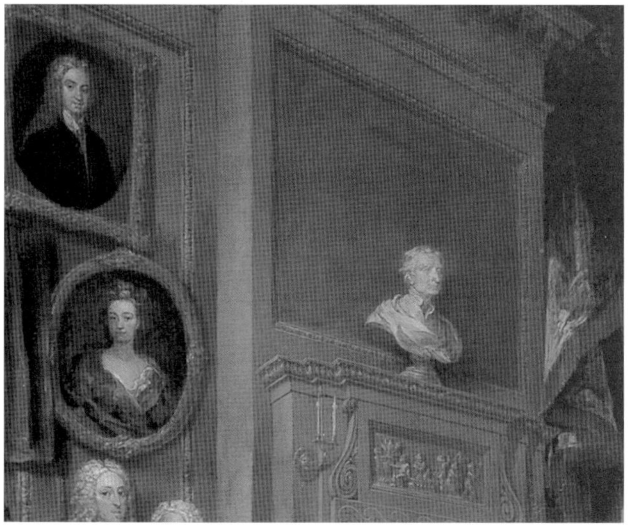

A.11
see colour Plate XIV

of Trade, with the natural philosopher John Theophilus Desaguliers, FRS (1683–1744), as the prompter at the side of the stage. On the wall to the left of the mantlepiece, on which the bust is placed, are portraits of John Conduitt and his wife, Newton's half-niece, Catherine (1679–1740); in the chimney piece below the bust, Hogarth has depicted Rysbrack's emblematic relief, with putti, of Newton's 'several Discoveries', newly installed on his Westminster Abbey tomb (**D.5**).

Malcolm Baker in a forthcoming book on Roubiliac and the eighteenth-century portrait bust notices that the bust depicted in **A.11** combines Rysbrack's collared shirt with Roubiliac's drapery, so that in painting it Hogarth must have had a knowledge of both Rysbrack's bust (**A.10**) and Roubiliac's bust (**A.19**). He also adduces that John Conduitt commissioned **A.19** as well as **A.10**.

A.12 Stone bust, after 'Conduitt bust' (**A.10**), *c.*1727–9.[12]
62.2 cm ($24\frac{1}{2}$ in) high, 55.5 cm (22 in) wide
Lord Cobham's Temple of British Worthies at Stowe, Buckinghamshire.

The bust, which has a repair to its nose following vandalism by two boys from Stowe School in 1990, was originally accompanied by an inscription in the stone above the alcove in which it is placed:

Sir Isaac Newton
whom
the GOD of Nature made to
comprehend all his Works;
and, from simple Principles, to
discover the Laws never known before,
and to explain the Appearances never understood,
of this Stupendous Universe.

but it now only has:

SIR ISAAC NEWTON
WHOM
THE GOD OF NATURE MADE TO
COMPREHEND HIS WORK.

Begun about 1726 and completed in 1729, James Gibbs (1682–1754) designed a building, at first called The Temple of the Worthies, then Gibbs's Building and finally The Belvedere, for Richard Temple, Viscount Cobham (1675–1749) in the gardens of his house at Stowe. This had eight stone portrait busts by Rysbrack standing on pedestals in a semicircle round its west end.[12] According to George Vertue (*Notebooks*, III: pp. 71, 133[28]) in 1732: 'Queen Elis. Ld. Verulam [Sir

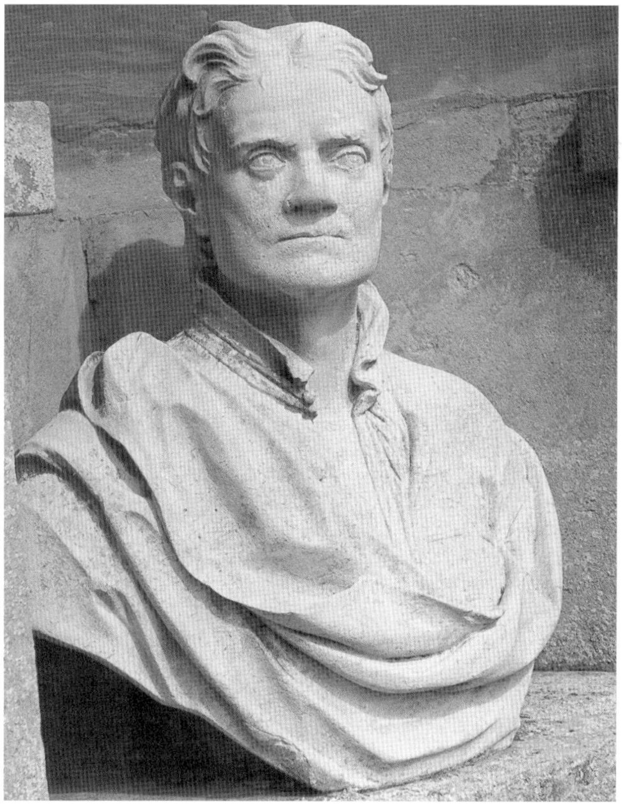

A.12
see colour Plate XV

Francis Bacon]. Shakespear. K. W^{m}. 3. Locke Newton Milton these busts made by M^{r}. Rysbrake in Temple of Worthies at Stowe.' – with the bust of Inigo Jones being the one omitted from this list.

The present Temple of British Worthies, to a design by William Kent (1684–1748) originally intended for Lord Burlington's gardens at Chiswick, was built *c.*1735 as a monument to British genius of those who were specifically 'Whiggish' heroes. It is placed some way from Gibbs's original building (later moved and rebuilt elsewhere in the gardens in 1764). The eight busts by Rysbrack were removed from their pedestals and placed with a bust of John Hampden, newly carved by him, in the sixteen niches of the new Temple, along with seven more busts, made in 1736–7 by Peter Scheemakers (1691–1771), of Alexander Pope, Sir Thomas Gresham, King Alfred, Edward the Black Prince, Sir Walter Raleigh, Sir Francis Drake and Sir John Barnard.[3,12]

A.13 Marble bust, after 'Conduitt bust' (**A.10**), but with round socle.
60 cm high, 56 cm wide ($23\frac{5}{8} \times 22$ in), socle 12.5 cm high (5 in)
Government Art Collection (GAC 16051) at 10 Downing Street, London, purchased in 1982 from Bruton Gallery, Somerset.

This bust is only doubtfully eighteenth century and possibly early nineteenth century.

A.14 Painted plaster bust after 'Conduitt bust' (**A.10**), (often wrongly called terracotta or painted terracotta), 1739.
60.5 cm high, 53.5 cm wide ($23\frac{3}{4} \times 21$ in)
Wren Library, Trinity College, Cambridge.[4]

Inscribed under right shoulder:

Sr ISAAC *NEWTON.*

and under left shoulder:

M. Rysbrack 1739

This bust was in Rysbrack's studio until 1756, when it was sold to Sir Edward Littleton (*c.*1726–1812), who bought eight more busts from him during the next ten years. It was with the Littleton family at Teddesley

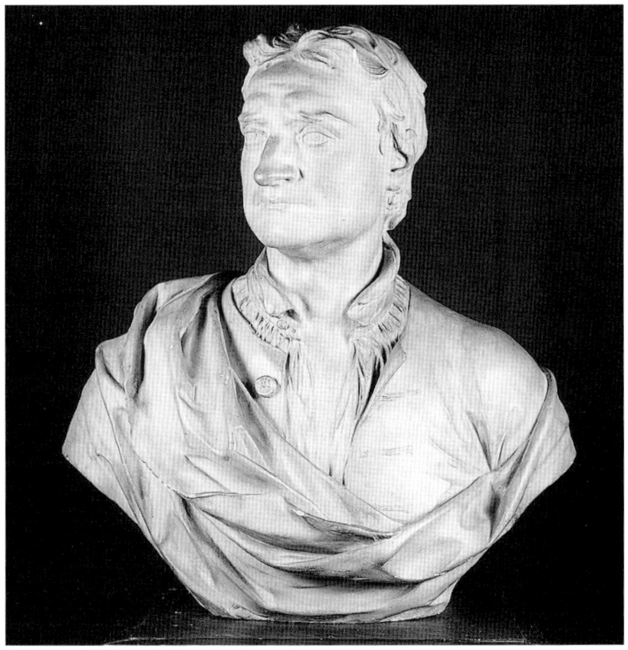

A.14

Park, near Stafford, until 1932 when it was exhibited at Spink's,[11] from where it was purchased by Sir Thomas Barlow (1883–1964) to give to Trinity College.

The College owns a letter to Sir Edward signed 'Mich: Rysbrack', and dated 'Febry 12. 1756': 'I have in the Whole four Busts finished; Milton. Sr Isaac Newton. Mr. Lock. and Bacon. and now I Promise you to Begin with the bust of Sir Walter Raleigh … The Reason of not sending Sir Francis Bacon is; that [as a terracotta bust] it must be Dried first, and Afterwards Burned which Cannot be Done till Summer, it not being half Dry Yet.' The Bacon was not dispatched to Teddesley Park till June 1757. The busts of John Milton and Alexander Pope, also in the 1932 sale, are now in the Fitzwilliam Museum.

Rysbrack did not do the casting of reproductions himself, but used Mr. Peter Vannini, 'the Caster in Plaster of Paris. (Whom I employ when I want)',[31] to make the piece moulds and the casts out of them, and it is possible that **A.14** is one of these, though signed and dated by Rysbrack, and long kept in his studio.

A.15 Bronze bust, after 'Conduitt bust' (**A.10**).
53.3 cm high, 55.9 cm wide (21 × 22 in), on a marble column with a white marble base, 106.7 cm (42 in) high

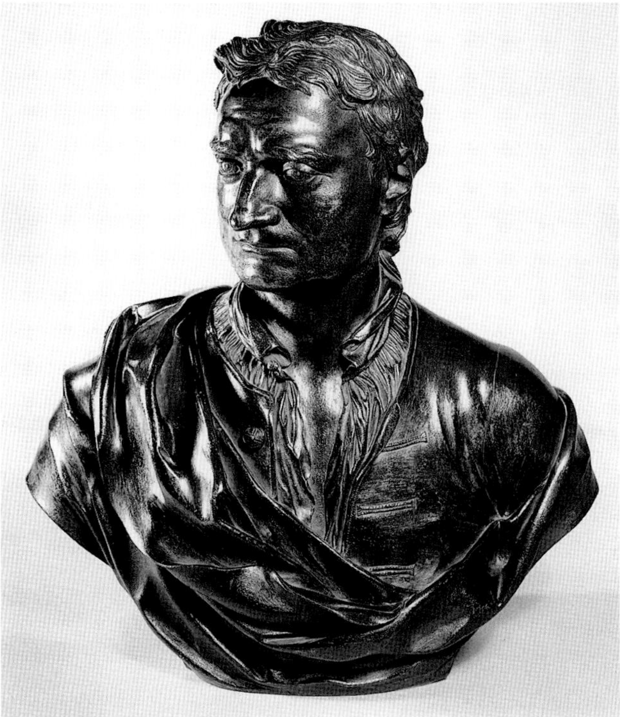

A.15

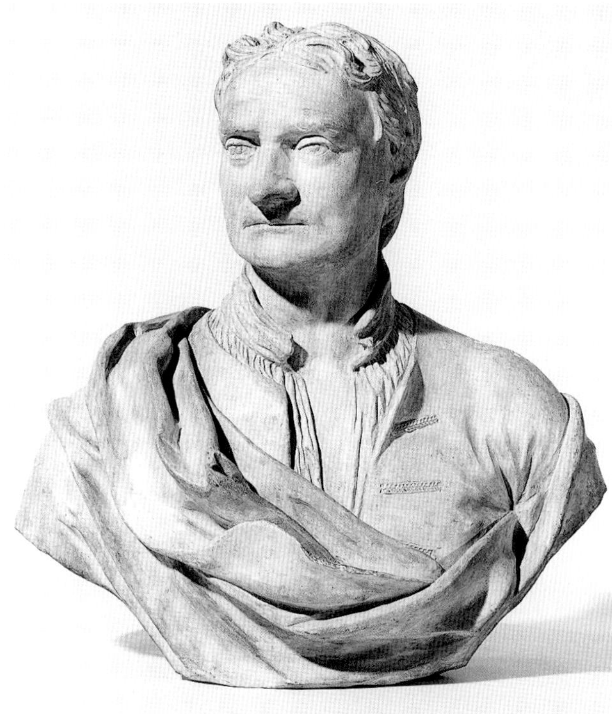

A.16

Trinity College, Cambridge; on loan to Fitzwilliam Museum from 2004.

Sold at Rysbrack's sale, 20 April 1765 (lot 75) as a bronze 'with a carved truss [support]'; in collection of O. V. Watney, Cornbury Park, Charlbury, Oxford; sold at Christie's on 22 May 1967 (lot 7) together with a bronze bust of Oliver Cromwell (lot 6), which had also been at the Rysbrack sale (lot 74), when bought by Humphrey Whitbread (1912–2000), who bequeathed it to Trinity College; on loan to Cecil Higgins Museum, Bedford, 1967–2004.

A.16 Plaster bust, after 'Conduitt bust' (**A.10**). 54.5 cm high, 52.8 cm wide ($20\frac{3}{4} \times 20\frac{1}{2}$ in) British Museum MLA SL 1985.[8].

Bequeathed by Sir Hans Sloane, FRS (1660–1753), and in Montague House, afterwards the British Museum, 1754. Sloane succeeded Newton as President of the Royal Society from 1727 to 1741. It is likely that this plaster bust is one of the casts made for Rysbrack by Peter Vannini. There is a pencil drawing of it, with drawings of busts of John Ray, FRS (1627–1705) and Isaac Barrow (1630–1677) by Roubiliac, by Sir George Scharf (1820–1895) in a notebook of 1881 in the National Portrait Gallery Archive (TSB XXVII, p. 52), reproduced in Dawson, *Portrait Sculpture*, 1999, p. 154, fig. 42.[8]

A.17 A bust of Newton is illustrated in a portrait of *Benjamin Franklin*, oil on canvas, 124.5 × 101.6 cm (49 × 40 in), by **David Martin (1737–1798)**, painted in 1766. The White House, Washington DC, since 1962.

After his testimony in the House of Commons in 1766 led to the repeal of the Stamp Act, Benjamin Franklin, FRS (1706–1790) was painted by Martin in London on the commission of Robert Alexander of Edinburgh. The bust in new bronze (painted plaster) is placed in profile on the left of the table, at which Franklin, wearing glasses, sits reading a document, with Newton's gaze directed towards him, invoking 'the greatest English voice of Reason'. Eight copies of the portrait were made in the eighteenth century, including three by Martin in 1767, 1772 and 1775, and two by Benjamin West (1738–1820) between 1784 and 1790. The bust of

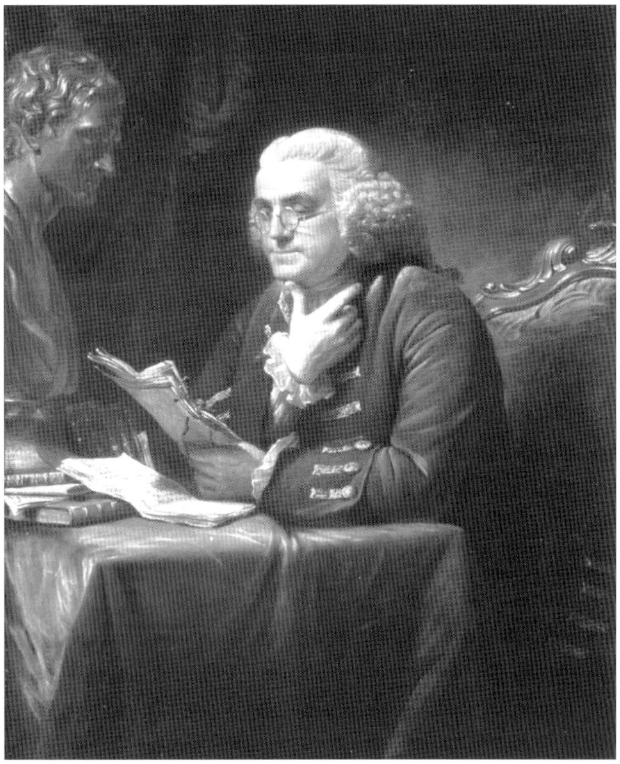

A.17
see colour Plate XVI

Newton may be after the Rysbrack 'Conduitt bust' (**A.10**), but is more possibly after the mezzotint (*c.*1736) by John Faber (**LIV-1**) of the marble bust (1732) by Guelfi in Kensington Palace (**A.9**), if the question of its being in new bronze is ignored.

Louis François Roubiliac (?1705–1762)

A.18 Terracotta bust with square socle.
58.7 × 49.2 cm (23 × 19⅖ in), socle 14.5 cm (5¾ in) high
Royal Greenwich Observatory: on loan to Fitzwilliam Museum, Cambridge, 1996–2001 and to National Maritime Museum from 2001.

Bust, head half-turned to left, wearing folded drapes *all'antica* over a fine chemise beneath.

From Minutes of Council of Royal Society, VII, pp. 230–1, 18 August 1785: 'Bequeathed to Royal Society by John Belchier [FRS (1706–85)] with intention it should be placed at the Observatory in Greenwich Park … N.B. This bust in Terra Cotta was made under the Eyes of Mr Conduitt and several of Sir Isaac's particular friends by Roubiliac, from many Pictures and other Busts, and esteemed more like than anything extant of Sir Isaac.'[10] Malcolm Baker considers that the bust of Newton painted in Hogarth's *A scene from 'The Indian Emperour' or The Conquest of Mexico by the Spaniards* (**A.11**), 1732–5, uses elements of this bust, which suggests that it was completed *c.*1732. **A.18** has been recognised since 1965[3] as a preliminary bust, or model, for the marble bust (**A.19**) at the Royal Society (as well as for **A.23**, **A.24** and **A.26**). Baker further suggests **A.19** was commissioned by John Conduitt, a commission that was to be important to Roubiliac to get him going in England, where he had arrived from France in 1730. The terracotta bust **A.20**, a replica of **A.18**, was presented by Roubiliac to Hogarth, but it is not known if this was before he painted **A.11**.

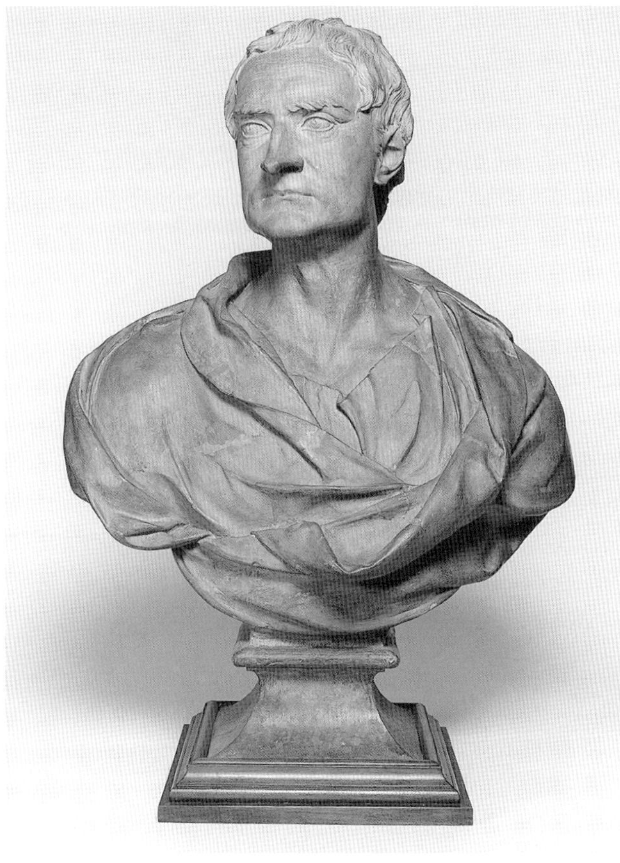

A.18

A.19 Marble bust, which follows **A.18**, with small square, decorated socle.
55.9 cm (22 in) high, with socle 12.7 cm (5 in) high
Royal Society, acquired in 1738.

Inscribed on socle:

NEWTON, 1738

The socle has a cartouche incorporating an astronomical diagram of the planets from *Principia* carved on it. As suggested by Malcolm Baker, it is conceivable that this earliest version 'of a fine marble Bust' was commissioned by, and executed for, John Conduitt, FRS (1688–1737), and sold after his death in 1737.[3] It was purchased by William Freman, FRS (d.1750) 'with an Intention of making a Present of it to the [Royal] Society' in April 1738 [*Journal Book*, vol. XVII, pp. 231–2, 13 April 1738].[3] As well, there is a Minute of Council (CMC vol. 2), for 19 June 1738: 'Mr Roubillac's Bill for a Pedestal to Sir Isaac Newton's Bust £2:7:0.' Esdaile[9] considered it was virtually a marble transcript (except for the classical drapery) of the portrait of Newton by Sir James Thornhill of 1709–11 (**IV-1**) in Trinity College, but also made using a copy of the death mask (**C.2**).

A.20 Painted plaster bust, a replica of **A.18**, missing since March 1864.

This bust was bought for National Portrait Gallery by Sir George Scharf (1820–1895), first Secretary 1857–82, then Director 1882–95, of NPG – see Secretary's Journal for March 1864 – at Christie's 17 March 1864 (lot 40). In the sale catalogue, was written:

> A Bust of Sir Isaac Newton, in terra-cotta, by Roubillac. This replica of the bust at Cambridge [**A.24**] was presented [? when] by the artist Roubillac to his friend, William Hogarth [1697–1764], as a token of esteem, and purchased from him by the celebrated John Hunter [FRS (1728–1793)], on the sale of whose effects Mr Sharon Turner [1768–1847] purchased it; in whose family it has remained ever since.

Hogarth, who was the son-in-law of Sir James Thornhill, the painter of three portraits (**IV-1**, **IV-2**, **IV-3**) of Newton, included a bust of Newton in a painting of 1732–5 of a theatrical scene in John Conduitt's house – see **A.11** – as well as in **A.21**. Whether **A.20** was given to Hogarth by Roubiliac before he painted **A.11** is not known; it was bought by John Hunter as lot 56 in *A Catalogue of the Pictures and Prints; the Property of the late Mrs Hogarth*, Greenwood, Leicester Square, on 24 April 1790 (lot 57 was Roubiliac's terracotta bust of Hogarth acquired by the National Portrait Gallery (no. 121) in 1861). Turner purchased it (lot 60) for two guineas on the third day of a four-day sale of Hunter's collection at Mr Christie's Great Room in Pall Mall on 31 January 1794 [James Christie (1730–1803)].

When the bust arrived at the NPG it was found to have been wrongly catalogued, and to be a solid bust of plaster hidden by a coating of paint. There has been no record of any further appearance in a sale after it was returned to Christie's.

A.21 A Roubiliac bust, most likely **A.20**, is illustrated in a portrait of Benjamin Hoadly, FRS (1706–1757), oil on canvas, 60.7 × 47.9 cm ($23\frac{7}{8} \times 18\frac{7}{8}$ in), painted *c.*1739 by **William Hogarth (1697–1764)**, in the Fitzwilliam Museum, Cambridge (no. 648) since 1908. There is a version of the portrait, dated 1740, in the National Gallery of Ireland in Dublin. Hoadly, elected FRS when he was aged twenty and Newton was still President, became an eminent physician, but why Hogarth placed the bust of Newton in his portrait is not known. Hoadly was the son of Benjamin Hoadly (1676–1761), Bishop of Salisbury 1723–34 and Bishop of Winchester 1734–61, who moved in the circle of Queen Caroline (1683–1737), particularly when she was Princess of Wales – see portrait of Dr Clarke **XVI-2**, with its portrait-profile of Newton.

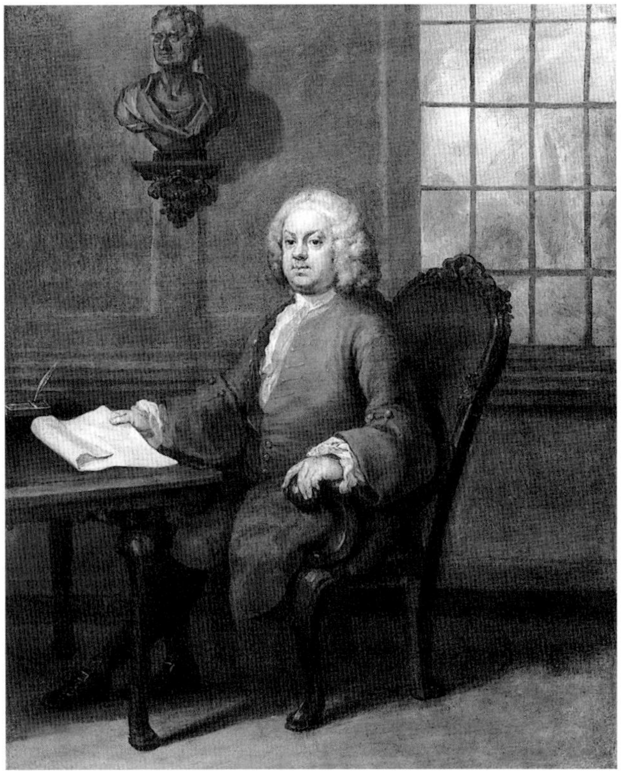

A.21

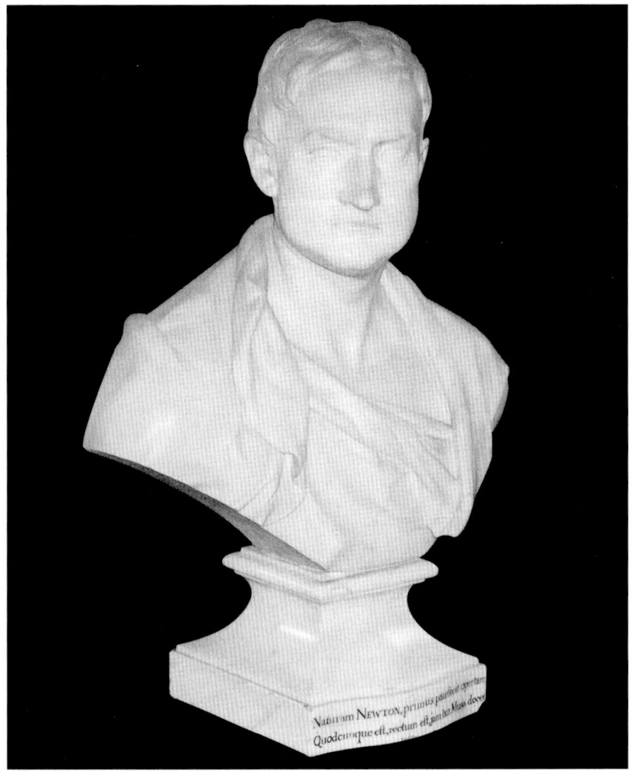

A.22
see colour Plate XVII

A.22 Marble bust, with a different drapery pattern from **A.18**.
54 cm ($21\frac{1}{4}$ in) high and socle 16.5 cm ($6\frac{1}{2}$ in) high
Lord (Jacob) Rothschild, London.

Inscribed on base of socle:

Naturam Newton, primus patefecit apertam.
Quodcunque est, rectum est, jam tua Musa docet.
[Newton was the first to reveal the secrets of nature.
Your Muse still teaches us: whatever is, is right.]

Provenance: 1st Earl Poulett, FRS (1663–1743), or, less likely, 2nd Earl (1708–1764), at Hinton House, Somerset, paired with Roubiliac bust of Alexander Pope, and then by descent; Sotheby's 1 November 1968 (lot 14); Cyril Humphris, 23 Old Bond Street, London; Benjamin Sonnenberg, New York City; P. & D. Colnaghi, 1979, to Lord Rothschild.

In Oliver Goldsmith's biography of Beau Nash of 1762[37] busts of Newton and Pope are mentioned as standing at the sides of a portrait of Nash by William Hoare on show in Bath – see **XVII-1**. It is tempting to suggest that they were the same two busts by Roubiliac owned by Earl Poulett 40 miles away from Bath at Hinton St George, and both in the sale of 1968.

A.23 Marble bust, which follows **A.18**, with square socle, in place by 1749.[3]
80 cm ($31\frac{1}{2}$ in) high including socle, unsigned
Library, Trinity College, Dublin.

In 1743, a sum of £500 was bequeathed by Dr Claudius Gilbert, the younger (1670–1743), Vice-Provost, 'for the purchase of busts of men eminent in learning to adorn the Library' of Trinity College, Dublin. The busts were taken on as a commission by Peter Scheemakers (1691–1771), but in part subcontracted to Roubiliac, according to Horace Walpole (1717–1797).[30] Malcolm Baker[3] has suggested that Roubiliac provided terracotta models for six of the busts, which were then cut in marble in the Scheemakers workshop, and that this bust resulted from a collaboration between the two sculptors.

A.24 Marble bust, which follows **A.18**, with square socle, 1751.
59.8 × 49.5 cm ($23\frac{1}{2}$ × $19\frac{1}{2}$ in), with socle 20.5 cm (8 in) high, 24 cm ($9\frac{1}{2}$ in) square at base
Wren Library, Trinity College, Cambridge.[4]

Inscribed:
(front of socle) *NEWTON*; (r.h. of support) *Ex Dono / Danielis Lock / Hujus Collegij A.M. 1751.*

Signed:
(l.h. of support) *L F ROUBILIAC / Sculp it. 1751.*

Daniel Lock, FRS (*c.*1682–1754) was a Fellow of Trinity College. The bust was said by Esdaile[9] to be 'virtually a marble transcript of Thornhill's portrait of Newton in the Master's Lodge' (**IV-1**), as well as use being made of the death mask.

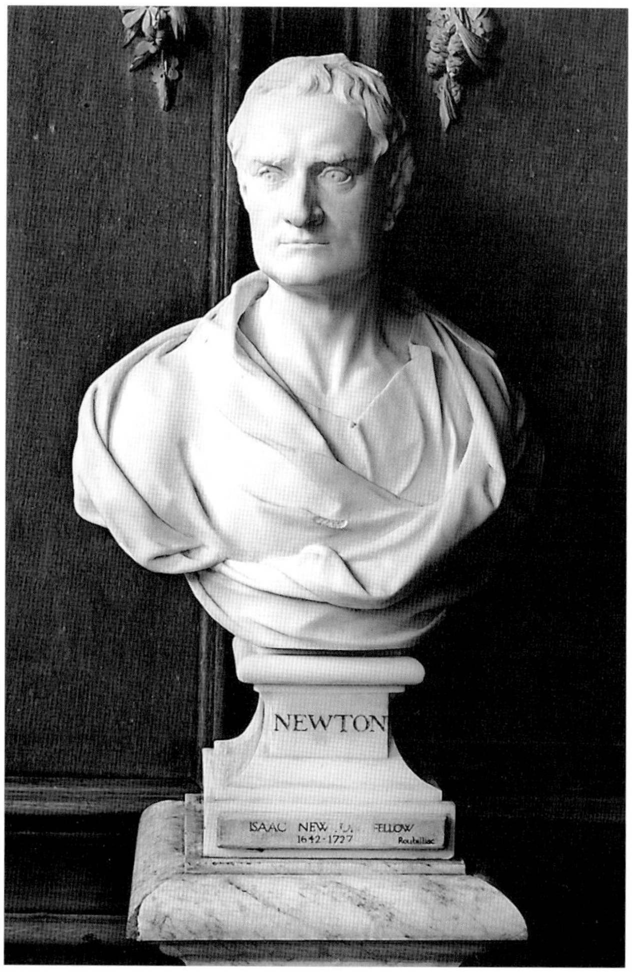

A.24

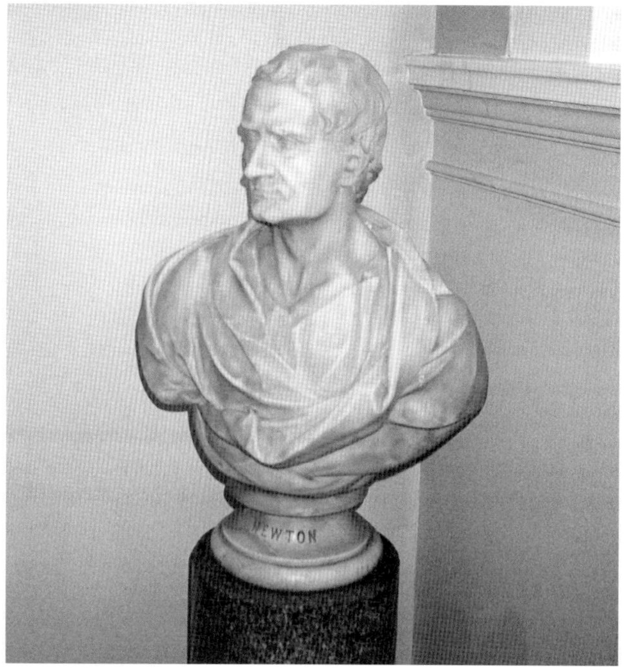

A.25
see colour Plate XVII

A.25 Marble bust, which follows **A.18**, late eighteenth century.
71.1 cm (28 in) high with round socle
Royal Astronomical Society, Burlington House, London.

A replica of this bust was reported by Chelminski, London, in 2003.

A.26 Plaster bust, which follows **A.18**.
58.7 × 49.2 cm (23 × $19\frac{2}{5}$ in), socle 14.5 cm ($5\frac{3}{4}$ in) high
Earl of Pembroke, Wilton House, Salisbury, Wiltshire, now missing.

It is recorded in James Kennedy's 'A new description of Wilton House' in 1779 as being paired with a terracotta bust of Sir Francis Bacon (1561–1626), also by Roubiliac, still there. It appears in an inventory (p. 42) taken in December 1827 with other busts placed over the bookcases in the Library, and Esdaile[9] noted in 1924 that 'there is a very fine cast of the Newton bust, possibly from the terracotta [**A.18**], in the Library at Wilton'.

The 8th Earl of Pembroke (1656–1733), who was PRS 1689–90 and pall-bearer at Newton's funeral at Westminster Abbey, could have acquired **A.26** for Wilton, though the 9th Earl (1693–1751), FRS in 1743, is perhaps more likely to have made the purchase – see **XV-1**.

A.27 Plaster bust, after **A.18**, with round socle.
62.2 cm ($24\frac{1}{2}$ in) high, with socle 8.9 cm ($3\frac{1}{2}$ in) high
University of Cambridge at Institute of Astronomy.

A.28 Bronze bust, after **A.18**, on moulded circular socle and later cylindrical pedestal and plinth, reddish-brown patina, eighteenth century.
18.7 cm ($7\frac{3}{8}$ in) high; 44.5 cm ($17\frac{1}{2}$ in) high overall
Sold Christie's 5 July 1994 (lot 121).

Joseph Nollekens (1737–1823)

A.29 Marble bust on rectangular base.
59.7 cm ($23\frac{1}{2}$ in) high, 53.3 cm (21 in) wide;
base 19.1 × 35.6 × 20.3 cm ($7\frac{1}{2}$ × 14 × 8 in)
Salisbury and South Wiltshire Museum.

On front of base is brass plaque with inscription on it:

This bust of Sir Isaac Newton, by Nollekens,
is presented to the Salisbury and South Wiltshire Museum
in affectionate remembrance of Richard Fowler M.D. F.R.S. F.S.A. &c, and,
Ann, his wife; Two of the original founders of this Museum;
by Theresa John Cornwallis West, of Newlands, Hants,
widow of Fredc Rich[d] West Esq. M.P. J.P. of Ruthin Castle, Denbighshire: August 23, 1881.

An entry in the Museum Record Book for 1881 says that the bust was given to the Museum on the understanding that in the event of the Museum being broken up or dispersed, it shall be transferred to the Mayor and Corporation of the City of New Sarum – Richard Fowler, FRS (1765–1863) was physician to the Salisbury Infirmary 1796–1847.

Joseph Nollekens was apprenticed to Peter Scheemakers (1691–1771), and worked in Rome 1760–70; if the ascription of the bust is correct, Nollekens made it very soon after he came back from Rome, dating it *c.*1771 (John Kenworthy-Browne, 2001). At Salisbury, the hair, though copious, has a flatness which is echoed on a Nollekens bust of Lord Newborough of 1772; compared with the Roubiliac bust **A.18** the drapery is somewhat classicised and arranged in flat, inert folds, and in proportion the head appears rather small.

Attributed to Peter Scheemakers (1691–1771)

A.30 White marble bust on square socle, with mended fracture across middle of neck, *c.*1730.
61 cm (24 in) high, 53.3 cm (21 in) wide, with socle 15.2 cm (6 in) high
Louth Naturalists', Antiquarian and Literary Society, Louth, Lincolnshire – at present in store in The Mansion House, Louth.

Head turned to right; hair short; wearing loose drapery folded particularly over the left shoulder, over a fine chemise. It is neither signed nor dated.

Provenance: From the 'Inventory of the Household Furniture … of David Garrick [1717–1779]', National Art Library MS 86 NN4 (ii)a, p. 9: Garrick's collection was sold by Mr Saunders at his Great Room, 'The Poets' Gallery', No. 39, Fleet Street on 23 April 1823 and nine following days, on the death of Mrs Garrick, when the bust, described as sculpted by Roubiliac, was bought by Charles Tennyson d'Eyncourt, MP (1784–1861) for furnishing his Gothic Revival house, Bayon's Manor, at Tealby, Lincs; given to the Mechanics' Institute at Louth on its establishment in 1835 (see Donations' Book, 11 April 1835); the Institute purchased and moved into The Mansion House, Louth, built as the Town Assembly Room about 1750, in 1853; on the closing of the Institute in 1956, the building, still with Newton's bust on a corner shelf where it had been since 1853, was given to the Louth Naturalists', Antiquarian and Literary Society, which continued to lease it to the Lincolnshire Library Service as a branch

RIGHT: **A.25** Marble bust of Newton after L. F. Roubiliac bust in Royal Society (**A.19**).

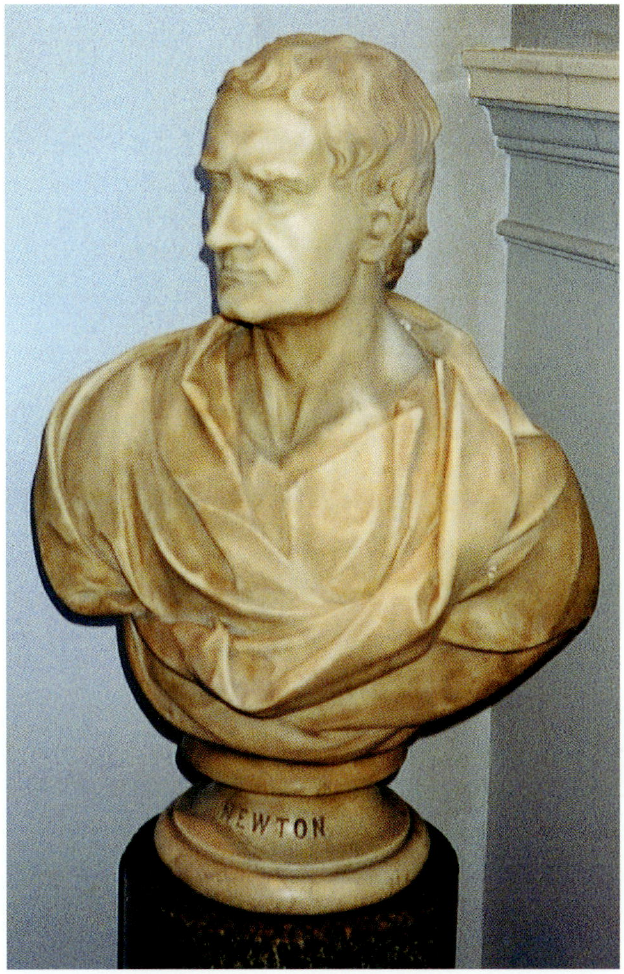

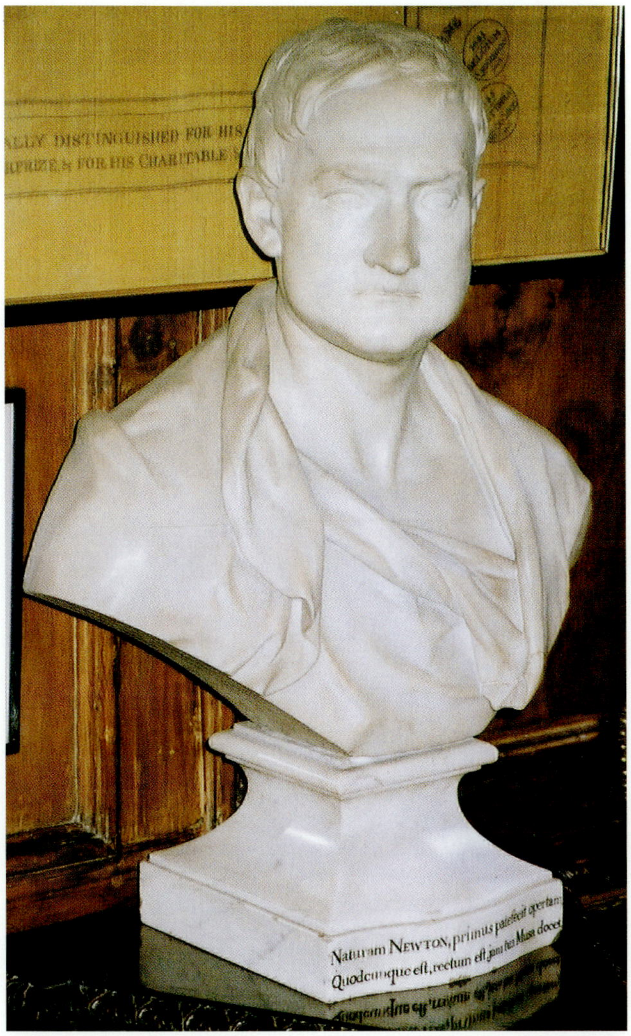

LEFT: **A.22** Marble bust of Newton by L. F. Roubiliac with a different drapery from **A.19**.

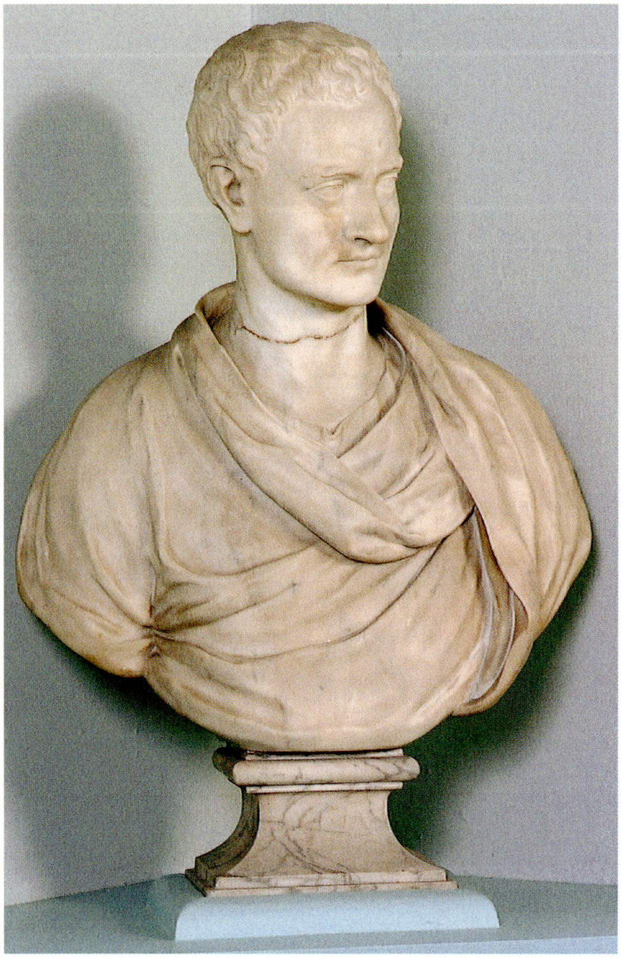

LEFT: **A.30** Marble bust of Newton attributed to Peter Scheemakers, c.1730.

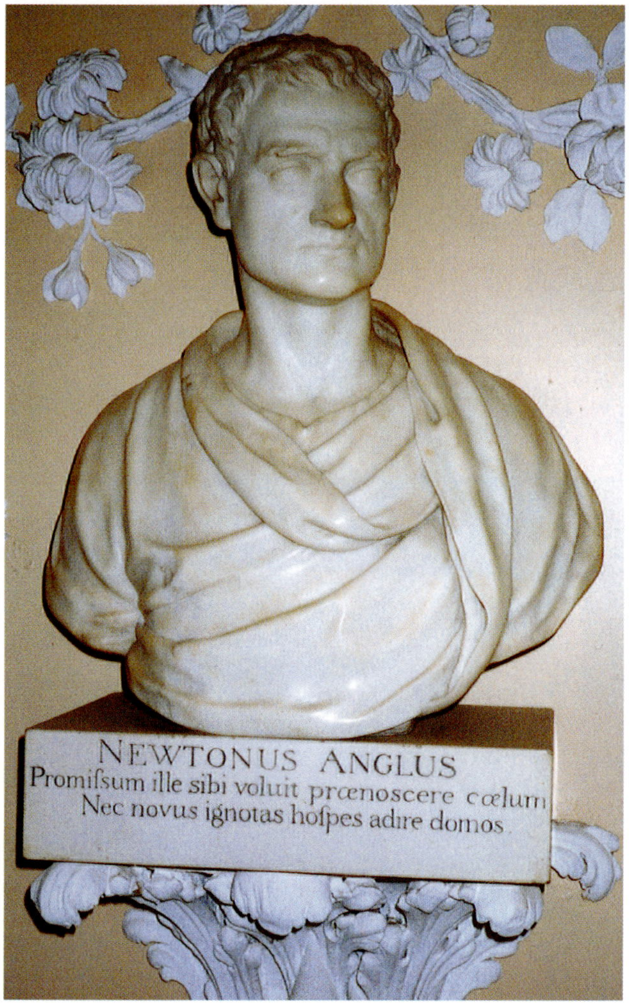

RIGHT: **A.31** Marble bust of Newton attributed to Peter Scheemakers, c.1760.

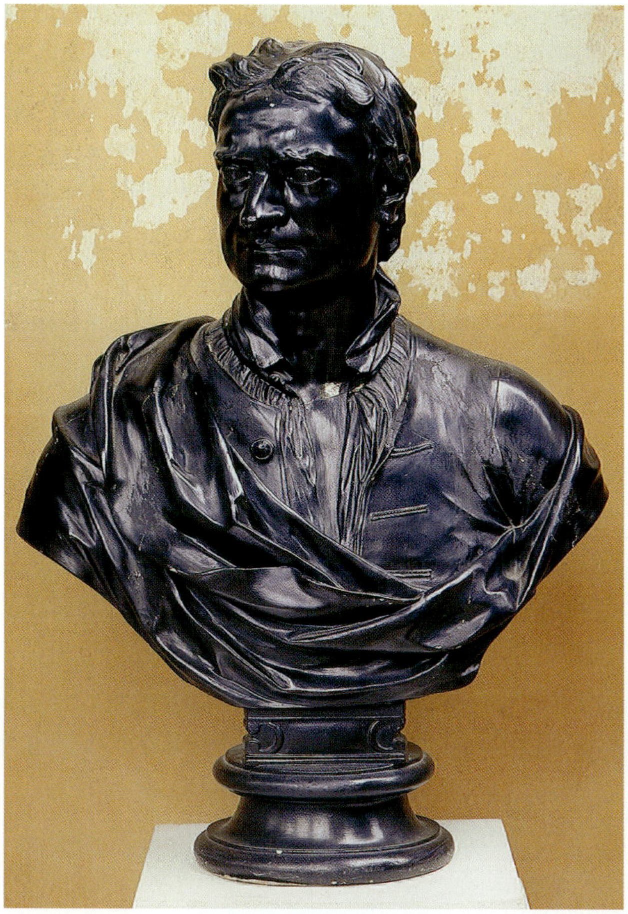

LEFT: **A.34** Black-patinated plaster bust, Type 1 after Rysbrack, by John Cheere.

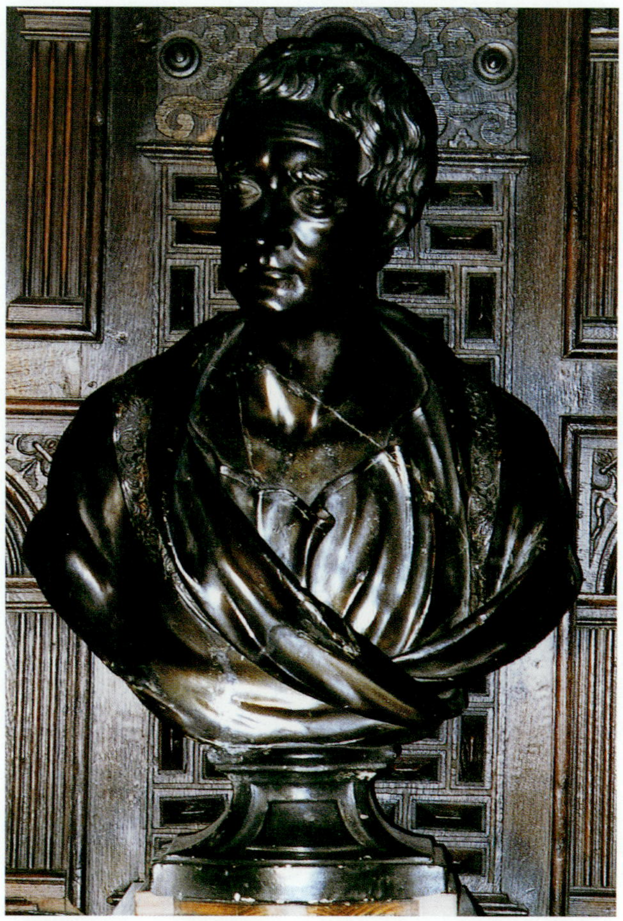

RIGHT: **A.38** Black-patinated plaster bust, Type 3 after Roubiliac, by John Cheere.

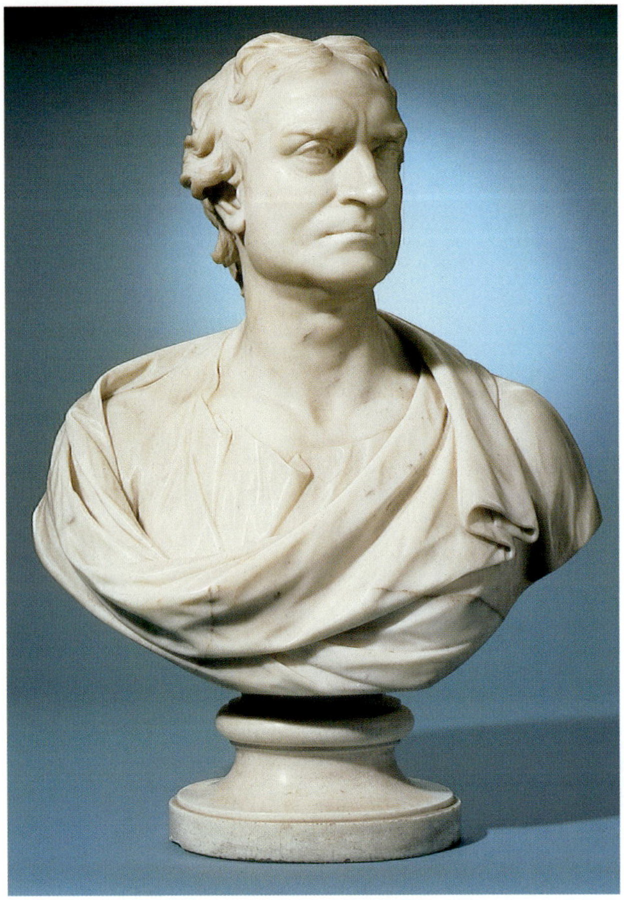

LEFT: **A.33** Marble bust of Newton by Joseph Wilton, 1762.

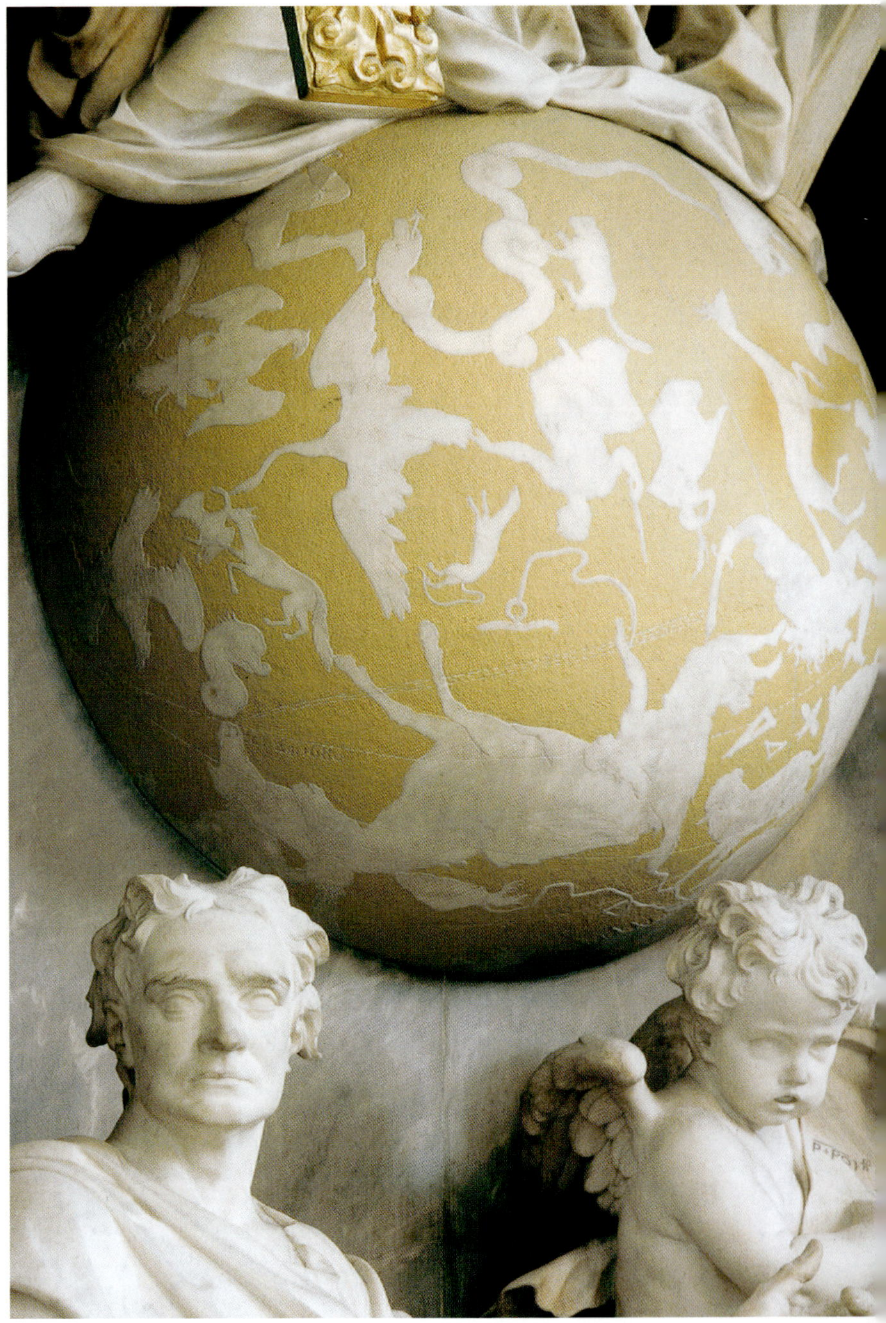

RIGHT: **D.5** Head and shoulders of Newton with one of the cherubs holding a scroll below the globe in the Monument in Westminster Abbey by Rysbrack, 1731.

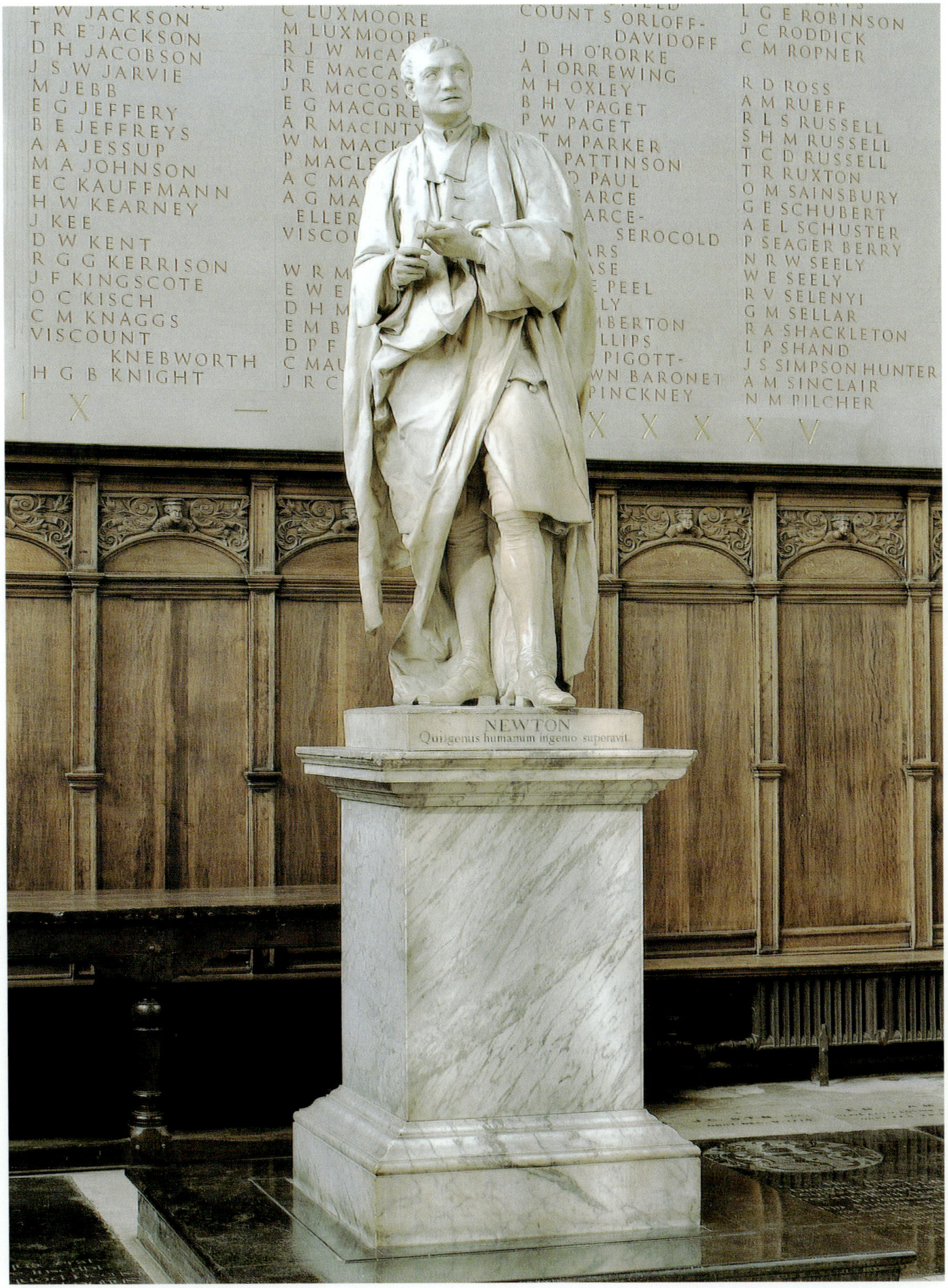

D.7 Marble statue by Roubiliac, 1755, in Ante-Chapel, Trinity College, Cambridge.

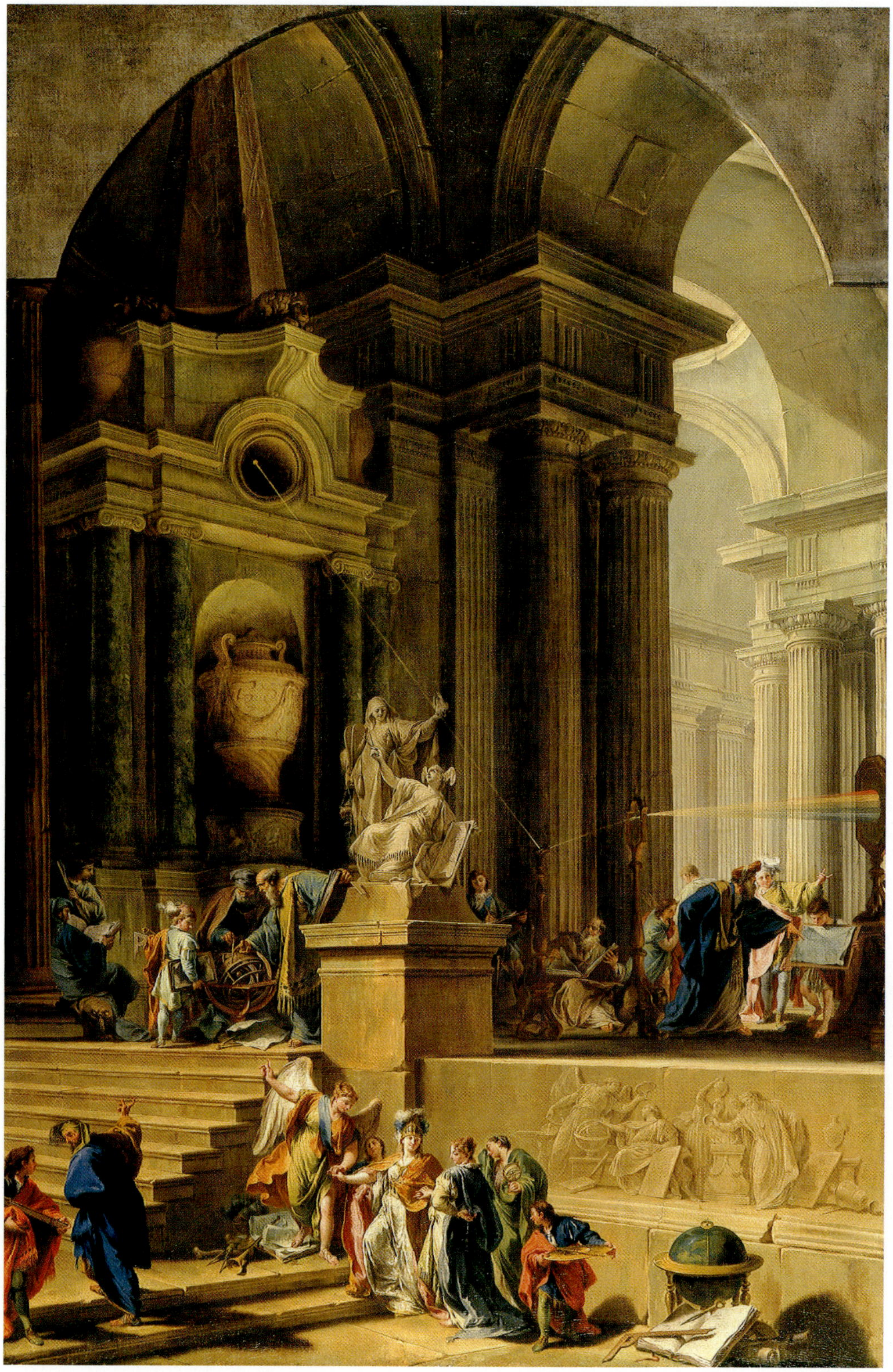

D.9 *An Allegorical Monument to Newton*, oil on canvas, by G. B. Pittoni with D. and G. Valeriani, 1730.

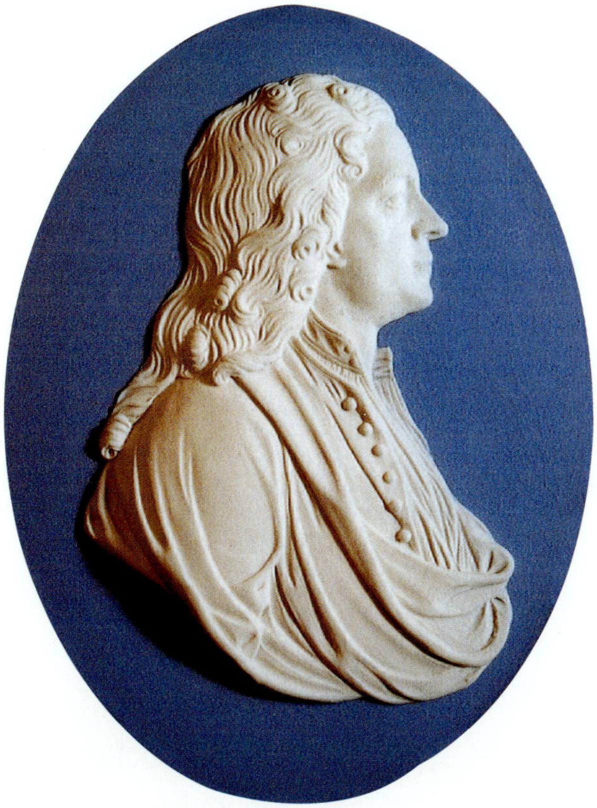

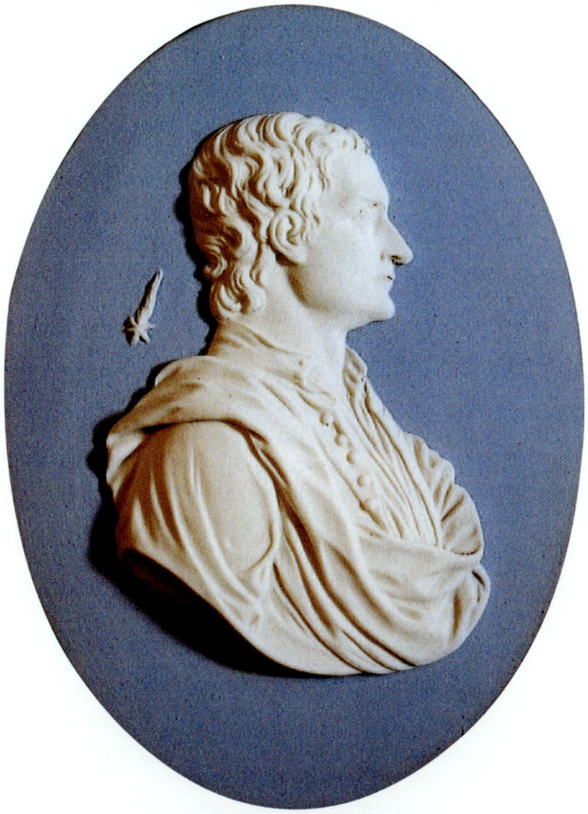

ABOVE: **E.2** and **E.3** Two Wedgwood portrait medallions of Newton, white on blue jasper.

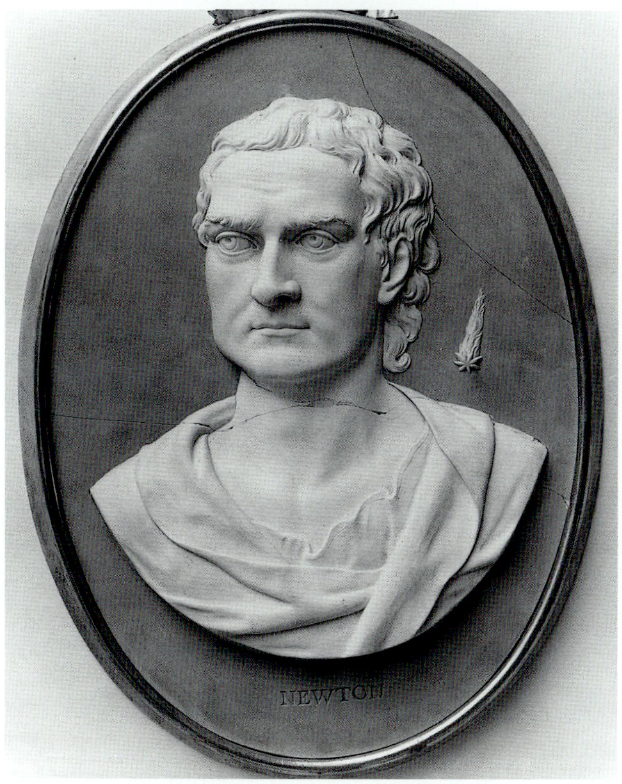

RIGHT: **E.7** Wedgwood portrait medallion, 1779–80.

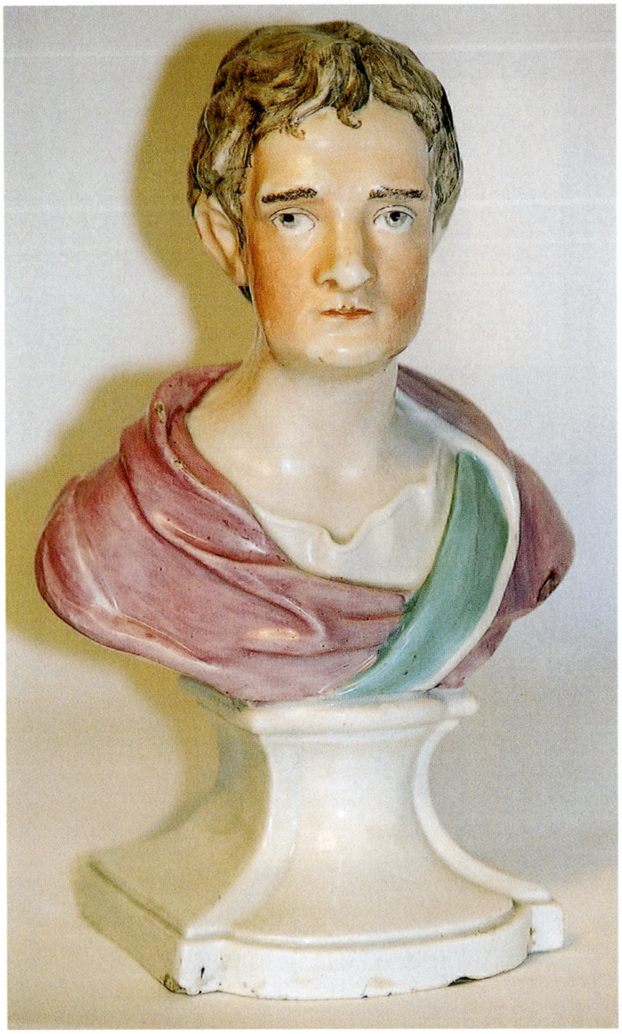

LEFT: **E.12** Ceramic bust, after John Cheere, by Enoch Wood.

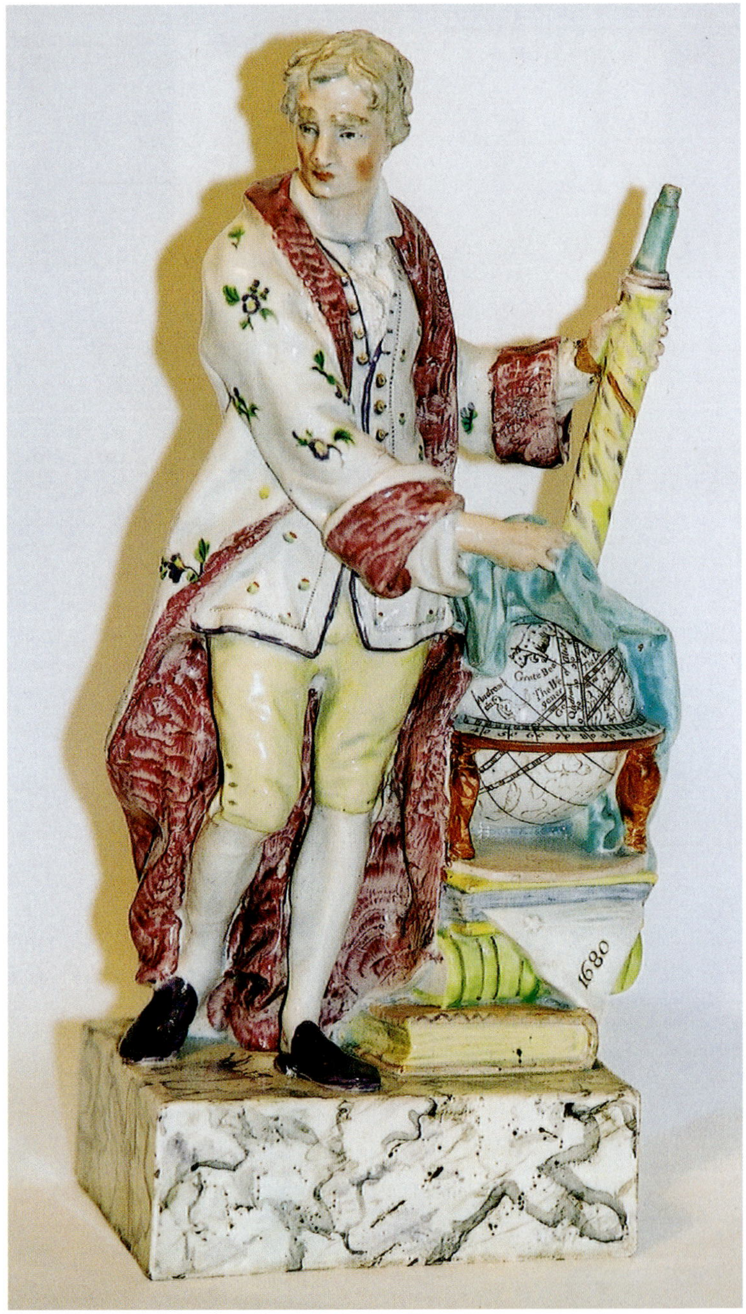

RIGHT: **E.14** Pottery figure by Ralph Wood, the younger.

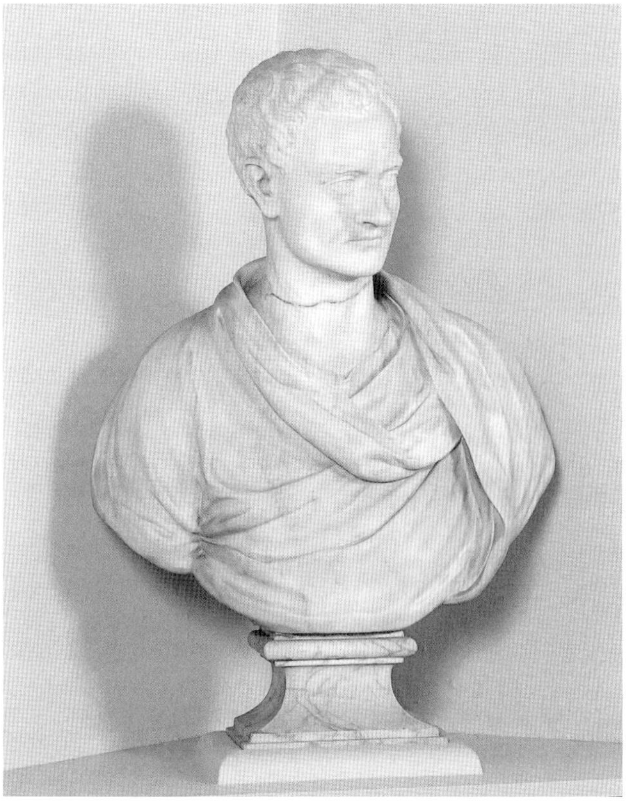

A.30
see colour Plate XVIII

library until 1991, when it became a store and second building to its museum.

Margaret Whinney (1897–1975), in a letter in the Conway Library to Rupert Gunnis (1899–1965) dated 10 December 1958, was certain the bust was not by Roubiliac, Rysbrack or Nollekens, but wondered about the Hendrik Scheemakers (1700–1748)/ Henry Cheere (1703–1781) studio. It is here attributed to Peter Scheemakers by Geoffrey Fisher (2001), who tentatively suggests a date of *c.*1730.

The early nineteenth-century attribution of the bust to Roubiliac may be explained by Garrick's friendship with the sculptor, who made two portraits of him in 1758: a profile medallion, now in the Garrick Club in London, and a terracotta, sold at the sale (lot 82) on 14 May 1762 of the studio contents at Roubiliac's 'dwelling-House in St Martin's Lane' by Mr Langford, from which a marble bust, now lost, and a plaster cast in the National Portrait Gallery (no. 707a) since 1883, were made. There is a painting by Andrea Soldi (*c.*1703–1771) of *Roubiliac modelling the bust of Garrick*, signed and dated 1758, at the Garrick Club since 1883. Soldi also painted portraits of Roubiliac (1751) and of Rysbrack (1753).

A.31 Marble bust on an oblong block, *c.*1760.
Bust 76 cm high, 61 cm wide (30 × 24 in), block 15 cm high, 45.7 cm wide (6 × 18 in)
Gothic Entrance Hall, Alscot Park, Stratford-on-Avon: private collection by descent.

Bust with head turned to right; hair short; wearing loose drapery folded particularly over the left shoulder, over a fine chemise. It is neither signed nor dated.

Inscribed on front of block:

NEWTONUS ANGLUS
Promissum ille sibi voluit praenoscere coelum,
Nec novus ignotas hospes adire domos.
['He wanted to find out in advance the nature of the heavens which would be his destination on dying, nor did he wish to arrive there as a guest in ignorance.']
Ovid, *Fasti*, 3, 159–60

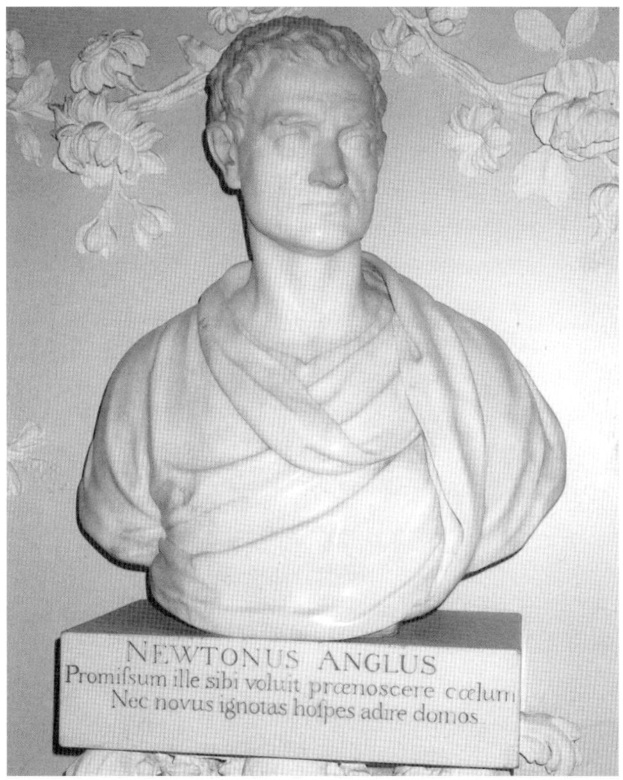

A.31
see colour Plate XVIII

James West, FRS (1704–1772), Joint-Secretary to the Treasury 1741–62, bought Alscot Park in 1749, and on retirement rebuilt and added to the house, completing it in 1764. In the Gothic Hall are a plaster bust of Shakespeare and marble busts of Newton and Matthew Prior. The bust of Shakespeare replaces the marble bust of him commissioned by West in 1758, signed by Rysbrack and dated 1760, but which only arrived at Alscot in 1763;[31] sold Christie's 15 July 1986 (lot 76); now in Birmingham City Museum and Art Gallery. Perhaps by their association with the bust of Shakespeare, both the Newton and the unattributed Prior bust have sometimes been wrongly labelled as in the style of Rysbrack, and neither is documented. Webb[31] stated without comment in 1954 that the Alscot bust of Newton was by Rysbrack, but it is here attributed to Peter Scheemakers by Geoffrey Fisher (2001) and dated *c.*1760. **A.31** is a version of the bust **A.30** at Louth made, it is thought, many years later.

There is a record of a plaster cast of Newton auctioned 6 June 1771, lot 21, in the sale of 'A Collection of models and casts of Mr. Peter Scheemaker' by Mr Langford – see Roscoe:[3] No. 315 within section on 'models and casts known only from documented sources', p. 301. And in the National Portrait Gallery Archive are photographs of a plaster bust, signed 'P. Scheemakers' and dated 1745, that was brought into the Gallery in 1965. The bust is possibly of Newton, wearing a three-buttoned open shirt under outer loose drapery, but with the nose showing signs of significant damage and repair.

Joseph Wilton (1722–1803)

A.32 Marble bust without socle, 1762.
50 cm tall, 52 cm wide, 29 cm deep ($19\frac{3}{4} \times 20\frac{1}{2} \times 11\frac{1}{2}$ in)
Acquired by the Bodleian Library, Oxford in 1762 (Bodleian Day Book for 1762);[20] at present on loan to History Faculty Library.

Head, with cropped hair, turned to right, loose drapery. Signed and dated 1762.

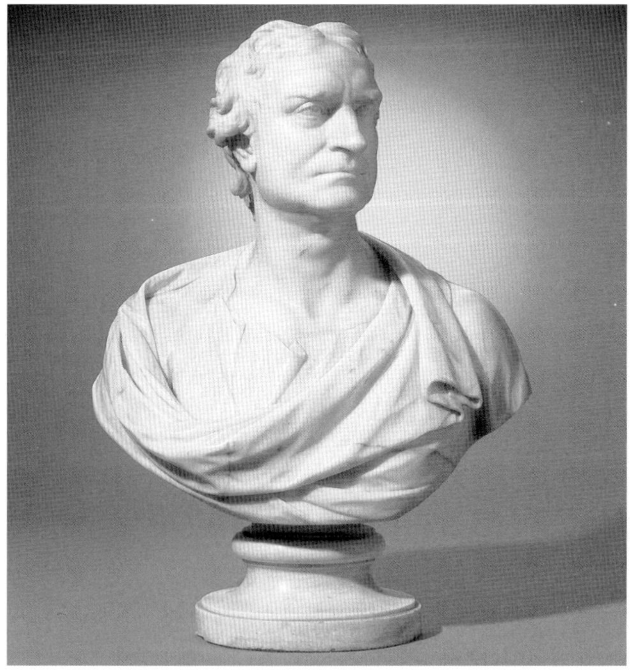

A.33
see colour Plate XX

A.33 White marble bust with oval socle, 1762.
71.1 cm (28 in) high including socle

Head, with short hair, turned to right; shoulders draped *all'antica*, a fine chemise beneath.

A version of bust **A.32** was recorded in a sale 'of a man of fashion' held by Mr Christie on 2 June 1779; Francis Broderip; Christie's 6 February 1872 (lot 108) catalogued as Roubiliac; Christie's 2 July 1991 (lot 81) when attributed to Wilton by Malcolm Baker.

John Cheere (1709–1787)[13]

John Cheere, the younger brother of Sir Henry Cheere (1703–1781) and a less able sculptor, had his manufactory for providing lead and plaster versions of antique statues and contemporary works at a yard adjoining Hyde Park Corner and Piccadilly. The plaster casts came in various sizes, and were frequently painted to imitate black-patinated bronze and sold in sets.

Another cast-maker in London was Charles Harris (d.1795), 'Opposite to the New Church in the Strand', who was also a maker of monuments. In his catalogue[13] of *c.*1790 there are plaster busts of Newton on pages 12, 13, 14, and 15 in four different sizes – 'large as life' (2 guineas), 24 inches (1 pound, 6 shillings), 16–18 inches (16 shillings) and 11 inches (7 shillings and sixpence).

Three types of busts of Newton by John Cheere have been identified: Type 1 after Rysbrack 'Conduitt bust' (**A.10**), and Types 2 and 3 after Roubiliac bust (**A.18**), which show a difference in the pronounced fold of the inner garment in the centre of the neck.

A.34 Type 1, after Rysbrack (**A.10**), black-patinated plaster bust.
75 cm ($29\frac{1}{2}$ in) high, 53.3 cm (21 in) wide, on tablet on round, shaped socle, together 16.5 cm ($6\frac{1}{2}$ in) high
Sir Edward Dashwood, Music Room, West Wycombe Park, Bucks.

West Wycombe Park was extensively added to and furnished from 1740 to 1780 by Sir Francis Dashwood, FRS (1708–1781), and contains three other black-patinated busts by John Cheere in the music room, those of John Milton, John Locke and John Dryden. There are many, though non-specific, payments to Cheere in the Dashwood papers.

A.35 Type 1, lead bust with rounded socle.
29.25 cm ($11\frac{1}{2}$ in) high
Paul Mellon Collection (B 1977.14.5), Yale Center for British Art, New Haven, Connecticut.

This bust has been attributed to John Cheere's older brother, Sir Henry Cheere, who is not known to have produced works in lead. Stylistically, when, for instance, compared with a lead bust of Dryden (19 cm ($7\frac{3}{8}$ in) high, Sotheby's, 12 April 1990, lot 134), it may safely be attributed to John Cheere. The bust of Dryden is also known in a larger bronzed plaster version.

A-36 Type 2, after Roubiliac (**A.18**), black-patinated plaster bust with rounded socle.
51 × 34 cm ($19\frac{7}{8}$ × $13\frac{3}{8}$ in) on socle 9.5 cm ($3\frac{3}{4}$ in), together 60.5 cm ($23\frac{3}{4}$ in) high.
Royal Institution of Great Britain, London.

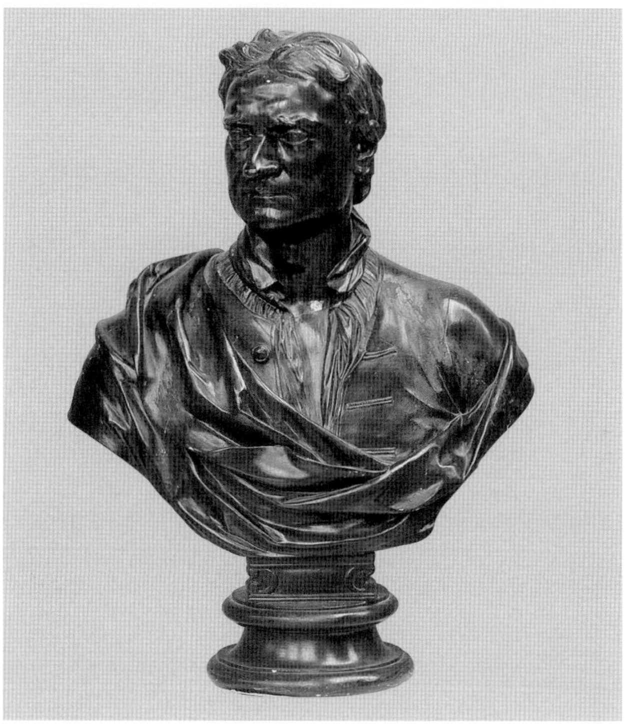

A.34
see colour Plate XIX

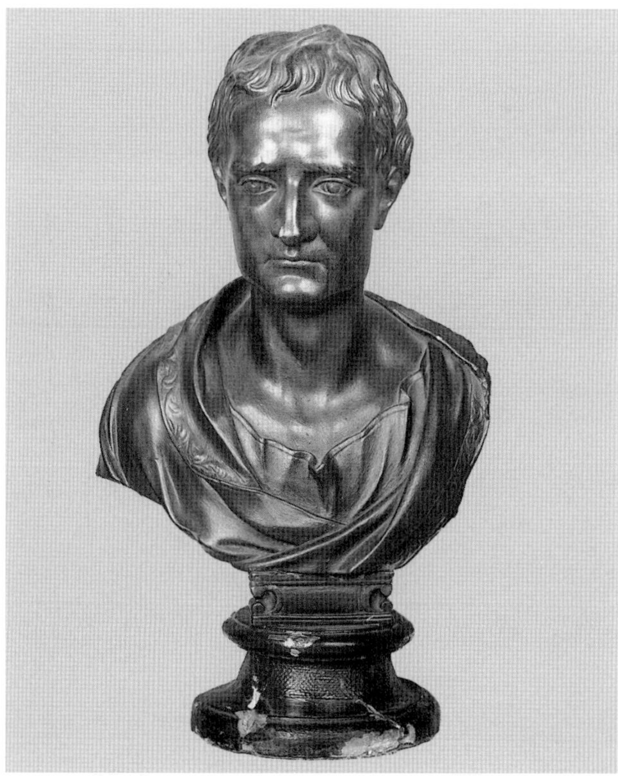

A.36

A.37 Type 2, white plaster bust with socle.
53.5 cm (21 in) high
Sir Edward Dashwood, Library, West Wycombe Park, Bucks.

The white plaster bust is one of five 'library' busts, all of the same height and with identical socles of a pattern commonly used by Cheere. The others represent John Milton, Joseph Addison, Alexander Pope and Socrates, and all are placed on the top of the bookcases.

A38 Type 3, after Roubiliac (**A.18**), black-patinated plaster bust on shaped socle.
71 cm (28 in) high, 48 cm (19 in) wide
Aston Hall, Birmingham since 1967.

At present displayed in the Long Gallery with five similar busts also placed on grained-pine 'Jacobean' plinths, copied from one by Richard Bridgens (*fl.*1811–1826): as well as the Newton, busts of Shakespeare, Milton, Pope, Locke and Inigo Jones were purchased from T. Crowther of Fulham in 1967. When James Watt, Junior (1769–1848) lived in the house between 1818 and 1848, he had there, besides one of Sir Isaac Newton, a group of busts of Oliver Cromwell, Dr Johnson, Inigo Jones, Horne Tooke, Thomas Telford, Sir Walter Scott, John Locke, Mrs Siddons, Francis Bacon and Shakespeare, sold at auction in 1849.

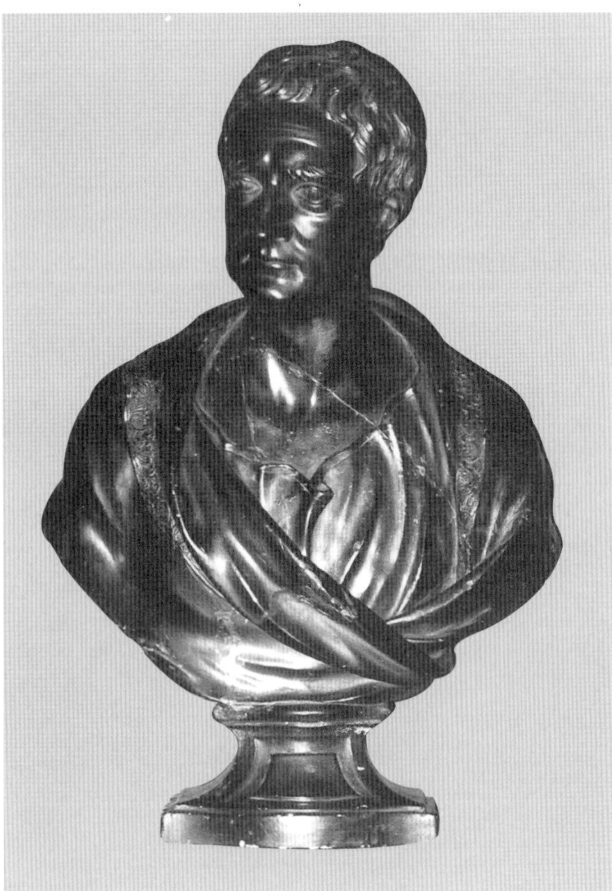

A.38
see colour Plate XIX

A.39 A white plaster bust of Newton and one of Shakespeare were ordered by the architect, John Carr (1723–1807), for 'the best staircase' at Fairfax House, Castlegate, York, in 1762, and supplied by John Cheere: 'Mr. Carr debtor to John Cheere to two large busts finished very neat and polished, of Sir Isaac Newton and Shakespeare two guineas each.'[13] They were placed on Corinthian brackets (still there) on the wall of the staircase hall, and remained *in situ* at least until about 1920 (A. Stratton, *The English Interior*, London: B. T. Batsford (1920), pp. xxviii, 43, pls. XCVIII, XCIX and fig. 37), but were found to be missing on photography for the National Monuments Record in 1942. From photographs used in Stratton's book the bust of Newton was John Cheere Type 3.

A.40 Type 3, bust of terracotta or artificial stone (?), on round socle.
76 cm (30 in) high
Private collection.

As John Cheere is not known to have produced works in either terracotta or artificial stone, it is likely that another manufacturer must have used his Type 3 mould. It has not proved possible to identify the material forming this bust.

B. RELIEFS

Attributed to Alexander van der Hagen (d.*c.*1779)[30]

B.4 to **B.10** are attributed to van der Hagen, of whom George Vertue (*Notebooks*, III: p. 135[28]) wrote in 1747: 'Mr Vander Hagen. Sculptor works for Mr Rysbrack. has done several heads portraits in Ivory.– very well. but not meeting with propper encouragement did not continue', and Horace Walpole (*Anecdotes of Painting in England*, 1771; vol. IV: p. 98[30]) that he 'carved heads in ivory'. He exhibited at the Free Society of Artists from 1766 to 1779.[30]

B.4 Ivory portrait relief, after Rysbrack bust **A.10**.
Oval, 11.7 × 9 cm ($4\frac{5}{8} \times 3\frac{1}{2}$ in)
Keynes Library, King's College, Cambridge.

Bust, with head in high relief slightly turned to the left, and own hair; wearing shirt and coat open in front; classical drapery over right shoulder and across chest.

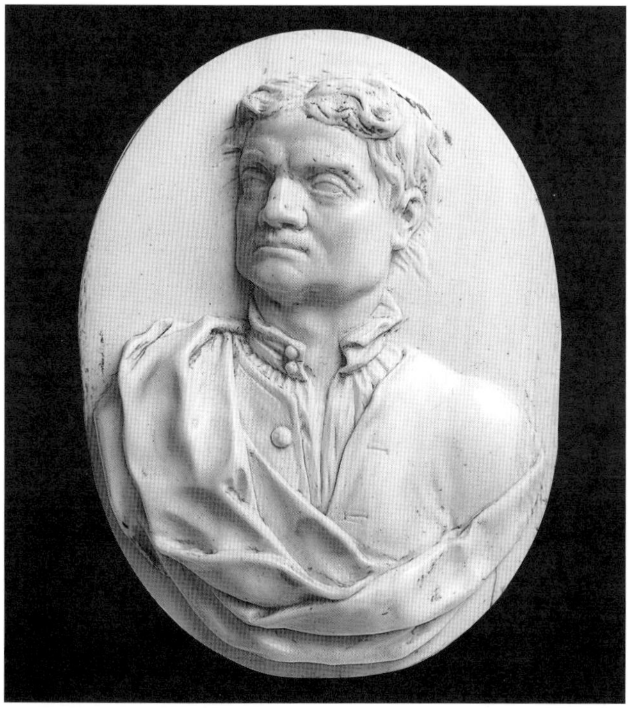
B.4

Inscribed (on back):
Isaac Newton / Eqa Aureus

Provenance: John Conduitt; by descent to Earl of Portsmouth; Viscount Lymington Sale, Sotheby's, 14 July 1936; then purchased from Dr Ernest Weil by John Maynard Keynes (1883–1946), and bequeathed by him to King's College, Cambridge, 1946.
Exhibited as by David le Marchand: *The European Fame of Isaac Newton*, Fitzwilliam Museum, Cambridge, 22 November 1973 – 6 January 1974;[6] *Footprints of the Lion: Isaac Newton at Work*, Cambridge University Library, 9 October 2001 – 23 March 2002.

B.5 Ivory portrait relief, oval, close to **B.4**.
11.5 × 9.3 cm ($4\frac{1}{2} \times 3\frac{5}{8}$ in)
Royal Society, London, purchased 1935 (Record of the Royal Society, 1935, p. 16).

B.6 Ivory portrait relief, oval, close to **B.4**.
13.5 × 10.3 cm ($5\frac{1}{4} \times 4\frac{1}{16}$ in)
Walters Art Gallery, Baltimore (no. 71.424), purchased before 1931.

The bust is based on Rysbrack's Conduitt bust **A.10**, but Charles Avery has suggested it could be derived from le Marchand portrait medallion **B.3**, now lost, but without the wig.

B.7 Ivory portrait relief, oval, close to **B.4**.
13 × 10 cm (5 in × $3\frac{7}{8}$ in)
Victoria and Albert Museum, London, no. A 112 – 1929; before purchase from Spero and Kevin, at Sotheby's, 20 June 1929 (lot 100).

B.8 Ivory oval plaque, profile to the left.
11.8 × 9.3 cm ($4\frac{5}{8}$ in × $3\frac{5}{8}$ in)
Grace K. Babson Collection (no. 498) in Burndy Library, Dibner Institute, Massachusetts Institute of Technology.

Mounted on wood, with a note of *c.*1900 on the back stating it is by Van der Hagen. The catalogue[2] of the

Babson Institute Library, 1950, notes that it is very similar to the portraits in ivory belonging to the Royal Society (**B.5**), and to King's College, Cambridge (**B.4**).

B.9 Black boxwood relief, bust close to **B.4**.
Oval 16.5 × 13.3 cm ($6\frac{1}{2} \times 5\frac{1}{4}$ in)
Paul Mellon Collection (B 1977.14.9), Yale Center for British Art, New Haven, Connecticut; bought at Sotheby's, 16 November 1972 (lot 88).

B.10 Pearwood plaque, oval, bust nearly full face.
Lost.
Sold Christie's, 5 June 1930.

Unattributed Profiles

B.11 Oval, fired clay, gilt high-relief of unwigged profile facing to right, with ground painted black and suspensory hole at top.

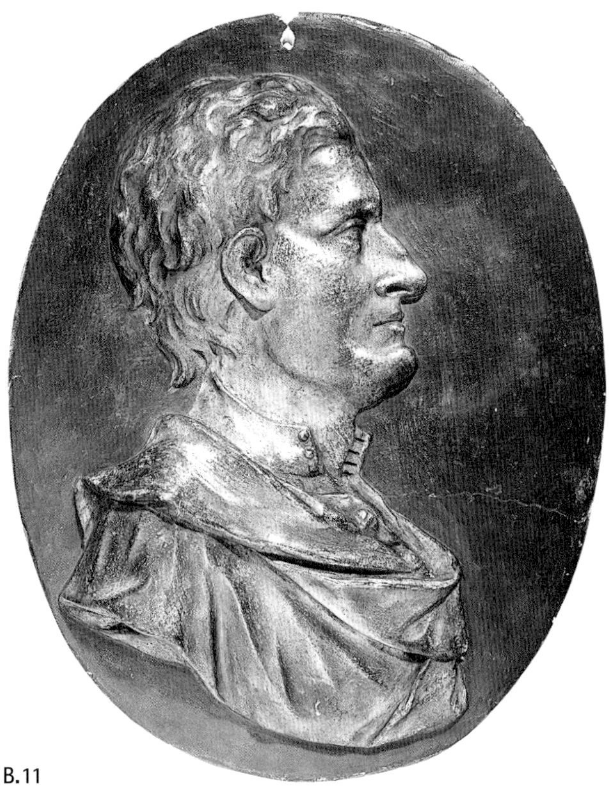

B.11

48.5 cm high, 30.4 cm wide ($19\frac{1}{8} \times 12$ in): relief 42.5 cm ($16\frac{1}{2}$ in) high, with its maximum width 25.5 cm (10 in); 2 cm thick.
British Museum MLA SL 1984[8] since 1754.

This portrait medallion, given to Sir Hans Sloane (1660–1753) before 1751 by the architectural writer Batty Langley (1696–1751), may have been made in Langley's factory, intended as an interior decoration. Discussed by K. Esdaile[10] in 1928.

B.12 Coloured wax, bas-relief profile bust.
11.2 × 8 cm ($4\frac{3}{8} \times 3\frac{1}{8}$ in)
Grace K. Babson Collection (no. 501) in Burndy Library, Dibner Institute, Massachusetts Institute of Technology.

John Jackson

B.13 and **B.14** Two turned circular medallions in wood, with relief profile portraits of Sir Isaac Newton, by John Jackson, 1749, which belonged to Sir Hans Sloane, FRS (1660–1753), and went to the British Museum with his collections in 1753 (MLA SL 2101).

One has a diameter of 6.1 cm and the other 4.5 cm. In each, Newton is facing left and has draped shoulders. Each inscribed:

ISAACVS NEWTON.

The two medallions were apparently both turned by John Jackson in an engine invented by him.

Louis François Roubiliac (?1705–1762)

B.15 Terracotta medallion, full-frontal bust.
Overall height, 28.5 cm; height of bust portrait, 24.2 cm
With Messrs Peel and Humphris Ltd in 1962.

This dates to the last ten years of Roubiliac's life.[10] On Roubiliac's death, at the auction sale of his studio contents at his 'dwelling-House in St Martin's Lane' by

Mr Langford on 12–15 May 1762, in the catalogue five 'plaister medals' of Newton are listed, of which Mallet[10] argues this is one, as well as moulds. There were also slightly smaller medallions in bronze. On the second day of the sale under the heading 'Bronzes, etc.' were two lots: '92 a basso relievo of Inigo Jones and Oliver Cromwell' and '93 Three ditto of Mr Handel, Sir Isaac Newton and Mr Pope'. And perhaps the same three 'basso-relievos' were in one of Mr Christie's earliest sales, held in either 1766 or 1767, as 'Sir Isaac Newton, Pope and Handel in bronze finely repaired [finished off] by the late ingenious Mr Robiliac'. The terracotta medallion is illustrated by Mallet,[10] and follows Roubiliac's marble bust **A.22** with its different drapery pattern from **A.18**.

Laurent Delvaux (1696–1778)[19]

B.16 Oval, bust-profile relief, bronze painted plaster (Jacobs, S 17).
35 cm (13¾ in) high
Musée Communal d'Archéologie, Nivelles, Belgium.

Newton looks to the left, with a descending comet placed behind his neck.
Signed with monogram: 'L.D.'

Provenance: Dr Le Bon (Fiévet); Société Archéologique de Nivelles.

This first attempt to create a portrait medallion of Newton is more severe and less idealised than the subsequent version. Above all his other discoveries, it was Newton's explanation of comets that particularly warranted glorification. Voltaire had written:

> Ces feux si long-temps inconnus, qui étoient la terreur du monde & l'écueil de la Philosophie ... sont mis enfin à leur véritable place par Newton.[29]

B.17 Oval, bust-profile relief, second version, in terracotta (Jacobs, S 21).
34.7 × 30.7 cm (13 11/16 × 12⅛ in)
Musées Royaux d'Art et d'Histoire, Brussels, since 1966.

Similar to **B.16**, in a classical style, with short hair and shoulders clothed in a toga. Not inscribed, signed or dated.
Provenance: Camille Bricourt (Fiévet); Albert de Wilde-Bricourt, Brussels.
This was, though not certainly, the model for **B.18**, **B.19** and **B.20**, which, as they have the year of Newton's death inscribed on them, can be dated after 1727.

B.18 Oval, profile relief after **B.17**, in marble (Jacobs, S 18).
37.5 × 33.5 cm (14¾ × 13 3/16 in)
Musée de Groesbeeck-de Croix, Namur, Belgium.

Inscribed:

ISAAC NEWTON ANGLUS NATUS 1642.
OBIIT 1727

Signed:

L. Delvaux

B.19 Oval, profile relief after **B.17**, in marble (Jacobs, S 19).
37.5 × 33.3 cm (14¾ × 13⅛ in)

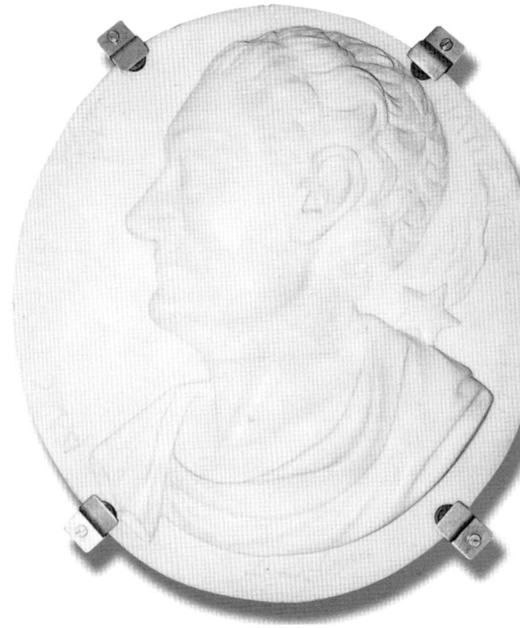

B.19

Spalding Gentlemen's Society Museum, Lincolnshire, since 1948 (no. B 130).

Inscribed:

ISAAC NEWTON ANGLUS NATUS 1642.
OBIIT 1727

Signed: 'L. Delvaux'

A 'Society of Gentlemen' was founded by the antiquary Maurice Johnson (1688–1755), at first in a coffee-house in the Abbey Yard, Spalding, in 1710, and then on a more permanent footing in 1712. It was established 'for the supporting of mutual benevolence, and their improvement in the liberal sciences and in polite learning'. Amongst the early members were Sir Isaac Newton, Sir Hans Sloane, Alexander Pope, George Vertue, Michael Rysbrack, William Stukeley, Richard Bentley and Robert Darwin (1682–1754), the father of Erasmus Darwin (1731–1802).

B.20 Oval, profile relief after **B.17**, in marble (Jacobs, S 20).
37.5 × 32.5 cm (14¾ × 12¾ in)
Collection: Paul Saintenoy, Ixelles; now missing.

Inscribed:

ISAAC NEWTON ANGLUS NATUS 1642.
OBIIT 1727

Signed: 'L. Delvaux'

B.21 Oval, profile relief after **B.17**, in plaster (Jacobs, S 22).
38 cm (15 in)
Provenance: Georges Fiévet, Bruxelles; now missing.

Inscribed:

ISAAC NEWTON ANGLUS

Not signed.

B.22 The same portrait medallion as **B.17 – B.21**, though in wood (Jacobs, SM 172), is placed over a door in a room of the Chapitre Noble de Sainte-Gertrude, Nivelles, with other eighteenth-century carvings.

D. WESTMINSTER ABBEY MONUMENT, TWO STATUES, ALLEGORICAL MONUMENT AND FRESCO, AND DESIGN FOR A CENOTAPH

Michael Rysbrack (1694–1770)

D.1 Preliminary design for the monument to Sir Isaac Newton, Westminster Abbey, red chalk, pen and ink and grey wash.[12]
32.9 × 23.5 cm (13 × 9¼ in)
Plymouth City Museum and Art Gallery.

Inscribed on the back in pencil (? copied from former mount):

For Sir Isaac Newton by M. Rysbrack

Provenance: Charles Rogers (1711–1784); 1784, William Cotton, by descent; presented to Plymouth Public Library in 1853.

This drawing reveals Rysbrack working out an entirely personal idea, unrelated to Kent's design.

D.2 Attributed to John Conduitt (1688–1737)

A drawing for a monument to Newton, pen and brown ink.[6]
22.5 × 17.5 cm (9 × 7 in)
Keynes Library, King's College, Cambridge since 1946.

Globe inscribed:

Isaacus Newton

Plinth inscribed:

Immortalem
Testantur, Tempus, Natura, Coelum
Mortalem
Hoc Marmor fatetur.
[Time, nature and the heavens bear witness to Newton's immortal genius; this monument acknowledges the man was mortal.]

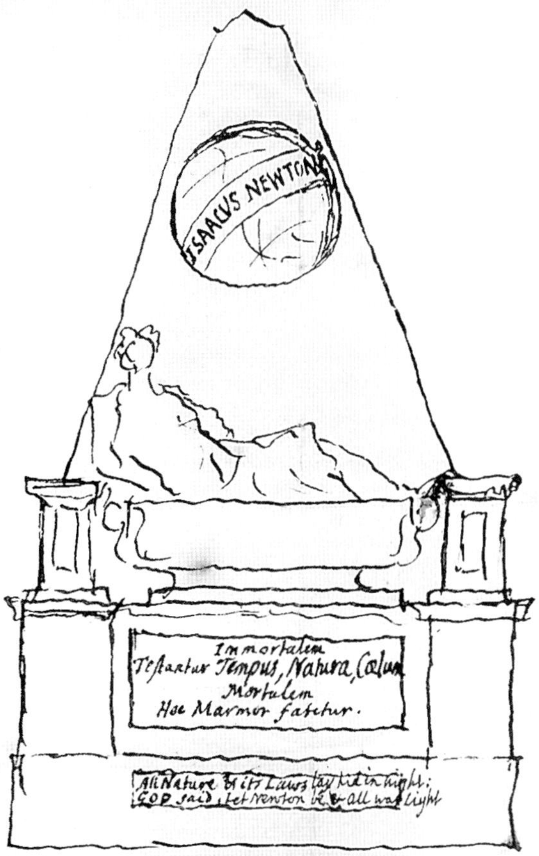

D.2

and inscribed on the base:

All Nature & its Laws lay hid in night:
GOD said, let Newton be & all was light.

By descent from John Conduitt in the collection of the Earls of Portsmouth; Viscount Lymington Sale, Sotheby's, 14 July 1936; collection of J. M. Keynes.

Conduitt and Pope were in correspondence about the monument in 1727. The drawing is presumed to have been made as a guide to Kent and Rysbrack. The words on the base, which were rejected, are an alternative version of Alexander Pope's famous lines:

> Nature and Nature's laws lay hid in Night:
> God said, let Newton be! and all was Light.

The drawing, with the reclining figure of Newton, the pyramid, the globe and the sarcophagus, contains all the essential features of the tomb as it was eventually made.

Conduitt outlined his ideas for the base-reliefs on the sarcophagus:

> In the mainground – A boy weighing in a stillyard [balance] the sun against all other planets – on one side – a boy looking up to the sky thro' a prism, another looking through a reflecting telescope … on the other side an aloes in a flower pot, to denote Vegitation and a Limbeck [distilling apparatus] to denote Chymistry …

The detail included a number of constellations engraved on the pyramid behind Newton, through which the path of the 1681 comet could be traced, as well as the 'solstitial colure' [a great circle of the celestial sphere passing through the poles of the equator and either the solstitial or the equinoctial points] shown at the point Newton presumed the Argonauts would have adopted.

The monument (1739) to John Conduitt in Westminster Abbey is the work of Sir Henry Cheere (1703–1781): the inscription on the plaque starts with a tribute to Newton. Cheere also made a bust of Conduitt's daughter, Kitty (1718–1750), who in 1740 married John Wallop (1718–1749), later Viscount Lymington when his father was created Earl of Portsmouth in 1743. The bust is now on loan from the Earl of Portsmouth to Winchester College.[13]

D.3 Michael Rysbrack (1694–1770)

Design for the monument to Sir Isaac Newton in Westminster Abbey, pen and brown ink, with brown wash.[6]
34 × 17.5 cm ($13\frac{3}{8} \times 6\frac{7}{8}$ in), with arched top
Department of Prints and Drawings, British Museum, purchased July 1859 (1859-7-9-99).

Signed at bottom, beyond margin:

Michael Rysbrack. Sculptor, fecit.

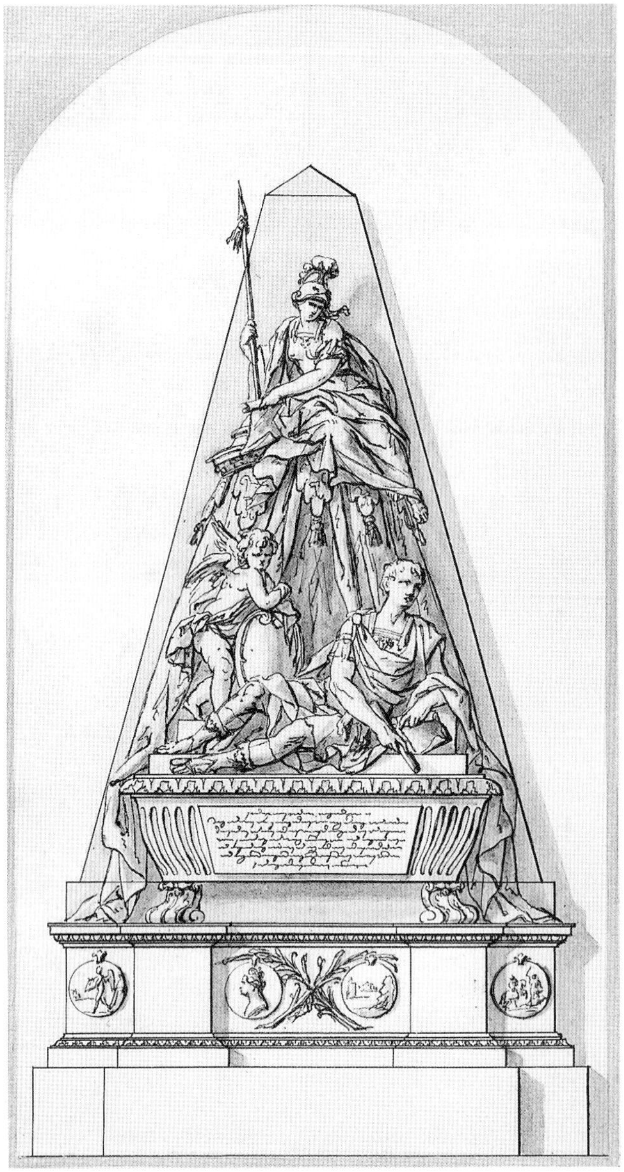

D.3

D.4 Michael Rysbrack (1694–1770) see page 100

A terracotta model by Rysbrack of Newton reclining on an oblong plinth, resting his right arm upon four volumes, his Mathematical Principia, his Optics, Chronology and Divinity – this last refers to his noble volume on the Prophecies of Daniel and the Apocalypse of St John.[22] His head is turned to the right, and his left hand raised.
Height 35.4 cm (13 15/16 in); length 52.9 cm (20 13/16 in); width 22.9 cm (9 in)

Victoria and Albert Museum (A.I – 1938).

The note in The *Gentleman's Magazine* of 22 April 1731 continues:

> On the Sarcophagus his own Figure is placed, in a cumbent Posture, his Elbow resting on the several Books written by him

And, though not shown with the model, the note continues:

> two Boys stand before him with a Scroll, on which is drawing a remarkable Diagram relating to the Solar System; and over that a converging series, an Invention which shews the utmost of human Understanding.

In the monument the arrangement of the pile of books, on which Newton is leaning, has been changed to give stronger support.

William Kent (1684–1748) and Michael Rysbrack (1694–1770)

D.5 The marble monument of Sir Isaac Newton in Westminster Abbey, ordered by John Conduitt (1688–1737) at a cost of £500, designed by the architect William Kent, who was paid £50 for the 'sketch design', carved by J. M. Rysbrack, and signed by both Kent and Rysbrack, was installed in April 1731.

The monument has a pyramidal structure with, from top to bottom, an eight-pointed star, the astronomic muse Urania with her right hand holding a *radius*, seated on a globe, Sir Isaac reclining on a sarcophagus, his right arm resting on a pile of four books, his left hand pointing to a scroll held by two cherubs, the sarcophagus decorated with a panel showing putti engaged in scientific experiments, and on a plinth of the monument, an inscription tablet.[22]

The symbolism of the tomb is explained in The *Gentleman's Magazine* of 22 April 1731:

> On a Pedestal is placed a Sarcophagus (or Stone Coffin) upon the Front of which are Boys in Basso-Relieve with instruments in their Hands, denoting his several Discoveries, viz. one with a Prism; on which principally his admirable Books of Light and Colours is founded; another with a reflecting telescope, whose great Advantages are now so well known; another Boy is weighing the Sun and Planets with a Stilliard, the Sun being near the Centre on one Side, and the Planets on the other, alluding to a celebrated Proposition in his Principia; another is busy about a Furnace, and two others (near him) are loaded with money as newly coined, intimating his office in the Mint.

Inscribed on the plinth of the monument are words composed by Richard Bentley (1662–1742) and Alexander Pope (1688–1744):

H. S. E.
Isaacus Newton Eques auratus,
Qui, animi vi prope divinâ,
Planetarum motus figuras,
Cometarum semitas Oceanique aestus,
Sua mathesi facem praeferente,
Primus demonstravit;
Radiorum lucis dissimilitudines,
Colorumque inde nascentium proprietates,
Quas nemo antea vel suspicatus erat, pervestigavit.
Naturae, Antiquitatis, Sacrae Scripturae,
Sedulus, sagax, fidus Interpres ;
Dei O. M. majestatem Philosophia asseruit,
Evangelii simplicitatem moribus expressit.
Sibi gratulentur mortales
Tale tantumque extitisse
Humani generis Decus.
Nat . XXV Dec. A. D. MDCXLII. Obiit XX Mar. MDCCXXVI

[Here lies entombed Sir Isaac Newton, Knight, who with the assistance of almost a divine inspiration of mind, and enlightened by the sun of his mathematical knowledge, first demonstrated the Motions and Figures of the Planets, the Paths of Comets, and the Causes of the Ocean's Tides. He thoroughly investigated the nice Distinctions of the Rays of Light, and the Properties of the Colours produced by them, of which no person before his time had entertained an idea. Of Nature, a diligent; of Antiquity, a sagacious; of the Holy Scriptures, a faithful Interpreter. He asserted the Majesty of Almighty God thro' the means of Philosophy, at the same time expressed the simplicity of the Gospel in his manners. Let mankind rejoice, on the existence of so great an ornament to the human race.[22]

The last three lines may also be translated:

> Let Mortals rejoice that there has existed such and so great an ornament of the Human Race.]

On the pavement before the tomb is a blue stone, on which is inscribed: 'Hic depositum est quod mortale fuit Isaaci Newtoni.'

George Vertue did not like the design of the monument, and wrote of it in his *Notebooks* (III: pp. 50–1):[28]

> April 1731. Sett up in Westminster Abbey the Monument of S^r^. Isaac Newton. a noble and Elegant work of M^r^. Michael Rysbrack. much to his Reputation. tho the design or drawing of it on paper was poor enough, yet for that only Mr Kent is honoured with his name on it (Pictor et Architect inventor.) which if it had been delivered to any other Sculptor besides Rysbrack, he might have been glad to have his name omitted.

The year of Newton's death on the monument, 20 March 1726, is given in Old Style. There is an allegorical poem on the subject of Newton's tomb by Orazio Arrighi Landini (1718–75) *Il Tempio della Filosofia … In cui s'illustrata Il Sepolcro d'Isaaco Newton*, Venice, 1755. First published in Florence in 1751, with second edition in 1752.

The Gothic Revival screen by Edward Blore, FRS (1787–1879) (**D.5b** and **D.5d**) gives a frame both to Rysbrack's monument to Newton and, on the other side of the entrance into the choir, to his monument to the Earl of Stanhope (1673–1721), also designed by Kent, and finished in 1733. It dates from 1834, and unfortunately, by obscuring part of each of the two

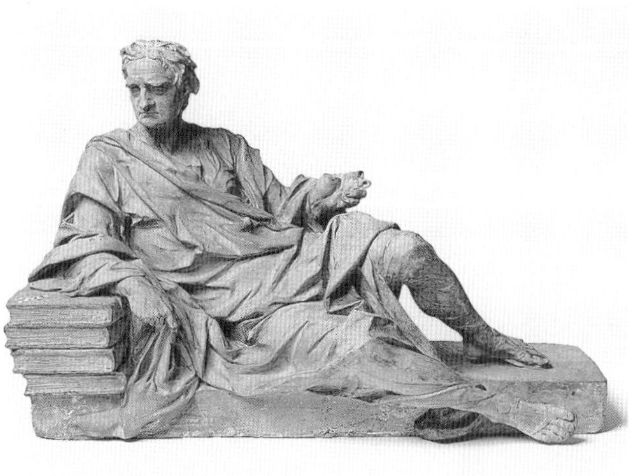

D.4

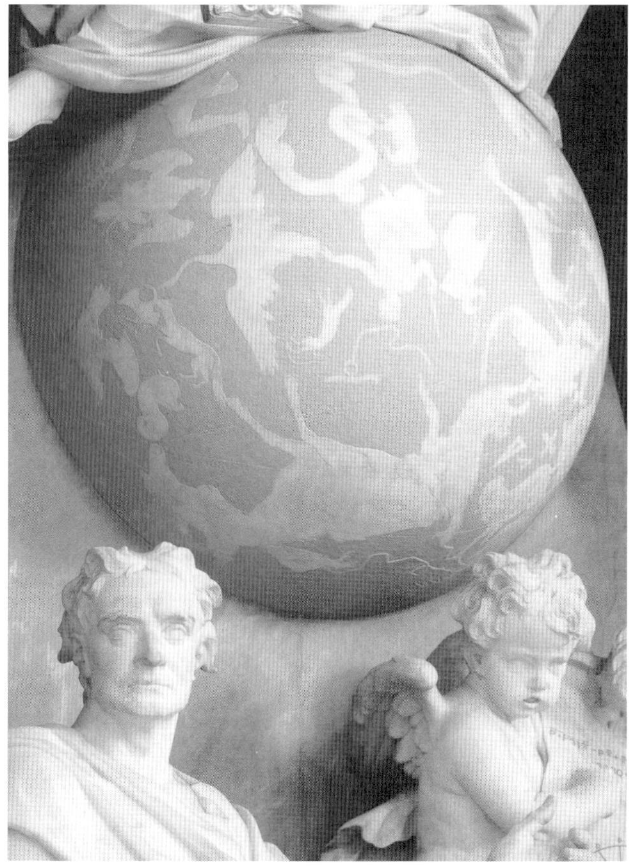

D.5c
see colour Plate XX

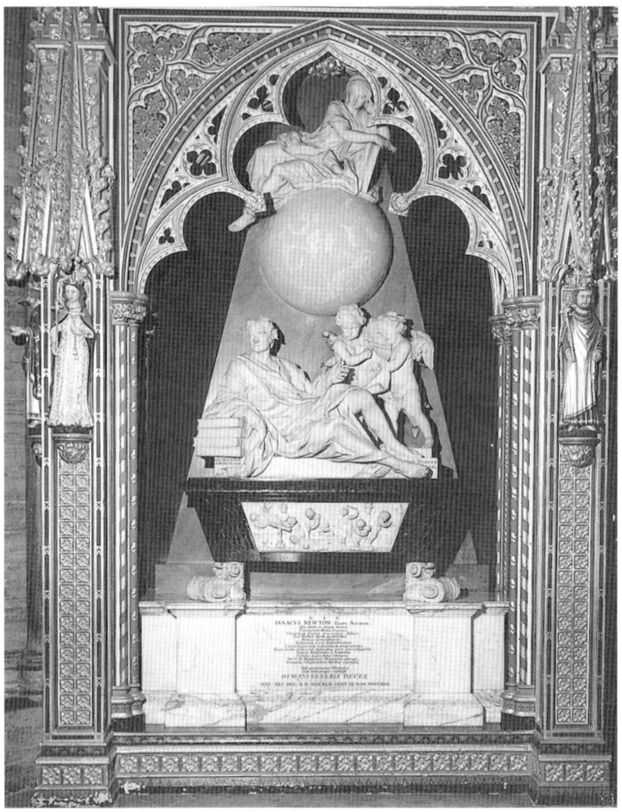

D.5b

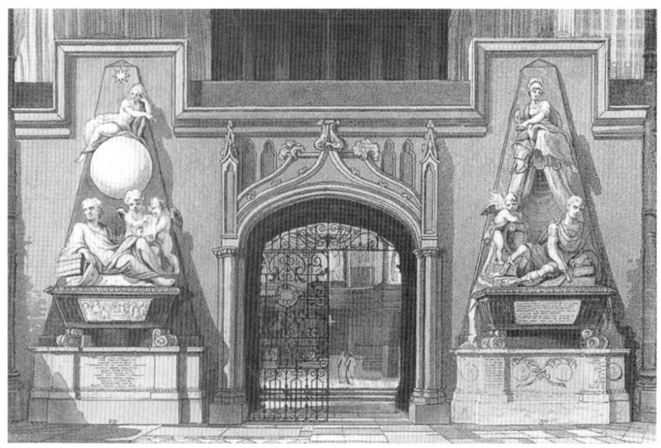

D.5d

monuments with its pseudo-Gothic tracery, distorts their classical proportions. Earlier, the choir screen designed by Nicholas Hawksmoor (1661–1736) (illustrated in *History of the Abbey Church of St Peter's, Westminster*, published by Rudolph Ackermann (1764–1834) in 2 vols. in 1812, drawn by Thomas Uwins (1782–1857), engraved by J. Bluck (*fl.*1791–1819)) had been plain, with no decoration other than the two monuments and the ironwork gate at the centre (**D.5d**).

See frontispiece **D.5a** for image of Newton in the monument.

D.6 Charles Reuben Ryley (1752–1798) after Kent and Rysbrack

Engraving, *c.*1736, of Monument erected to the Memory of Sir Isaac Newton in Westminster Abbey (**LIII-1**). 'Ryley del. published by R. Bowyer, Historic Gallery, Pall Mall, May 20 1794. Fittler Sculp. [Engraved by James Fittler (1758–1835)]'
28 × 21.2 cm (11 × $8\frac{3}{8}$ in)

LIII-1

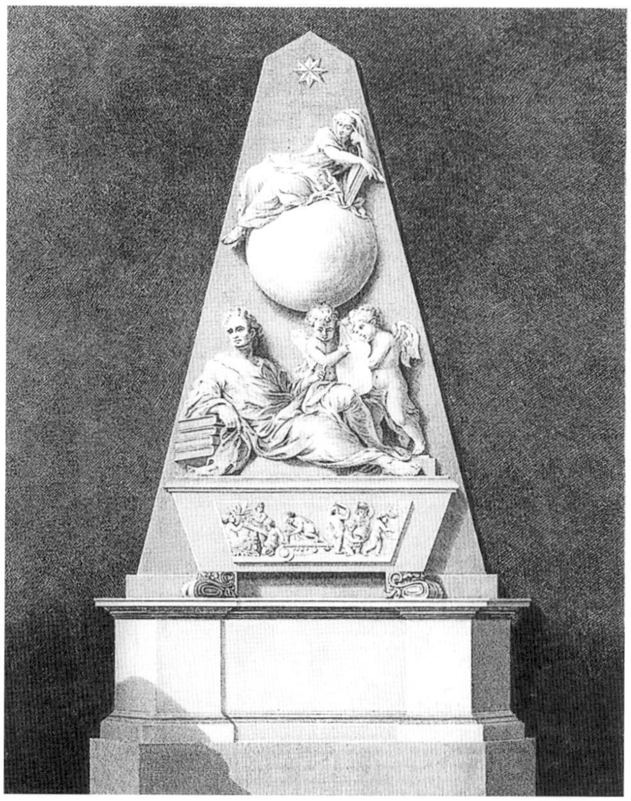

The engraving (**LIII-1**) does not show the inscription on the monument.

Louis François Roubiliac (?1705–1762)

D.7 Marble statue in Ante-Chapel, Trinity College, Cambridge, 1755.

The figure holds a prism in his right hand. It is 180.3 cm (71 in) tall, stands on a slab 63 cm ($24\frac{3}{4}$ in) square and 10 cm (4 in) high, placed on a marble base 83.8 cm (33 in) square and 120.6 cm ($47\frac{1}{2}$ in) high.

Inscribed at front of slab:

NEWTON.
qui genus humanum ingenio superavit.
['He surpassed the race of men in understanding' – said by Lucretius of the philosopher Epicurus]

and on back of slab on left:
Posuit Robertus Smith S.T. P. Collegij hujus S.Trinitatis Magister. MDCCLV.

Signed (on right at back):
L. F. ROUBILIAC inv^{it}. et sc^{it}.

Commissioned by Robert Smith, FRS (1689–1768), Master 1742–68, at a cost of £500. First mentioned in a letter from Edward Montagu (1691–1775) to his wife, Elizabeth (1720–1800), in July 1753.[10] Esdaile[9] found that the statue closely followed after Thornhill's portrait (**IV-1**) in Trinity, as well as using the death mask (**C.2**).

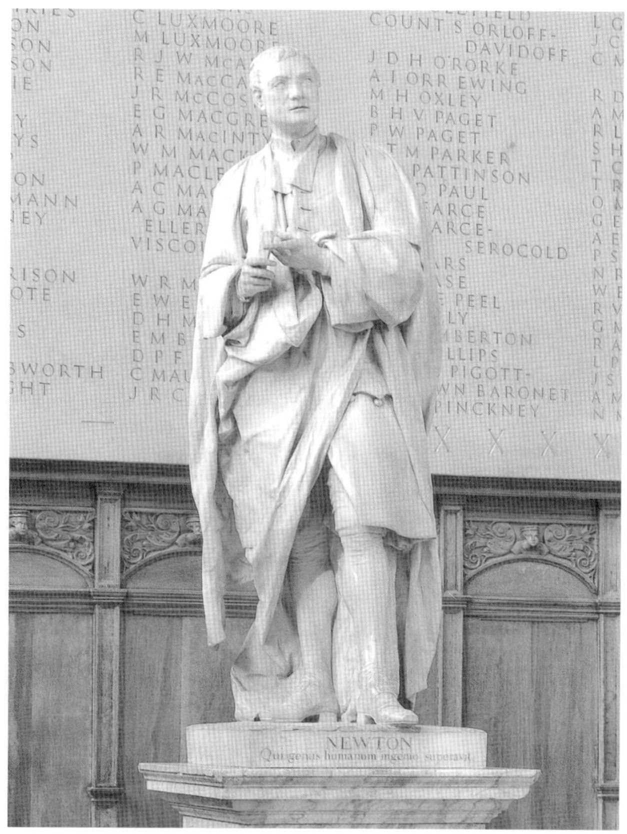

D.7
see colour Plate XXI

William Wordsworth (1770–1850) in his autobiographical poem, *The Prelude* (Book III, lines 60–3), was inspired by apparently being able to see the statue from his undergraduate rooms in St John's College:

> *The antechapel where the statue stood*
> *Of Newton with his prism and silent face,*
> *The marble index of a mind for ever*
> *Voyaging through strange seas of thought, alone.*

John Cheere (1709–1787)

D.8 Statuette, bronzed plaster, 1749.[13]
49.5 cm ($19\frac{1}{2}$ in) high
Transferred to York City Art Gallery in 1973 from York Castle Museum, to which it had come, after purchase in 1950, from Sir William Turner's Hospital at Kirkleatham House, Redcar, Yorkshire.

Newton standing with crossed legs besides a pedestal, on which is a pile of books on which he leans with his left elbow and with his left hand to his forehead; wearing a loose garment over an open-necked shirt with his right hand writing on paper beside the pile of books; on the side of the pedestal is a relief of Astronomy seated on a globe incised with constellations taken from the apex group in the Westminster Abbey monument.

Signed and dated on pedestal:
Cheere Ft 1749
and inscribed:
NEWTON

Giovanni Battista Pittoni (1687–1767) with Domenico and Giuseppe Valeriani (d.1761 and ?1708–1762, respectively)

D.9 *An Allegorical Monument to Sir Isaac Newton*, oil on canvas, finished by 1730.[6]
220 × 139 cm ($86\frac{5}{8} \times 54\frac{6}{8}$ in)
Fitzwilliam Museum, Cambridge (PD. 52 – 1973); on loan to Trinity College, Cambridge, since 2001.

Provenance: Collection of Sir William Morice; Capt. Heatley, by whom given to Ushaw College, Durham, *c.*1820; Sotheby's 21 March 1973 (lot 29).

One of a series of twenty-four paintings representing imaginary monuments 'of the British Monarchs, the valiant Commanders, and other illustrious Personages, who flourish'd in England about the End of the seventeenth, and the Beginning of the eighteenth Centuries.' The series was commissioned by Owen MacSwinny (before 1684–1754) from leading Venetian and Bolognese artists in Venice during the period 1720–30. The subject of the painting was discussed with John Conduitt in November 1727, June 1729 and September 1730, but remains a noble but generalised

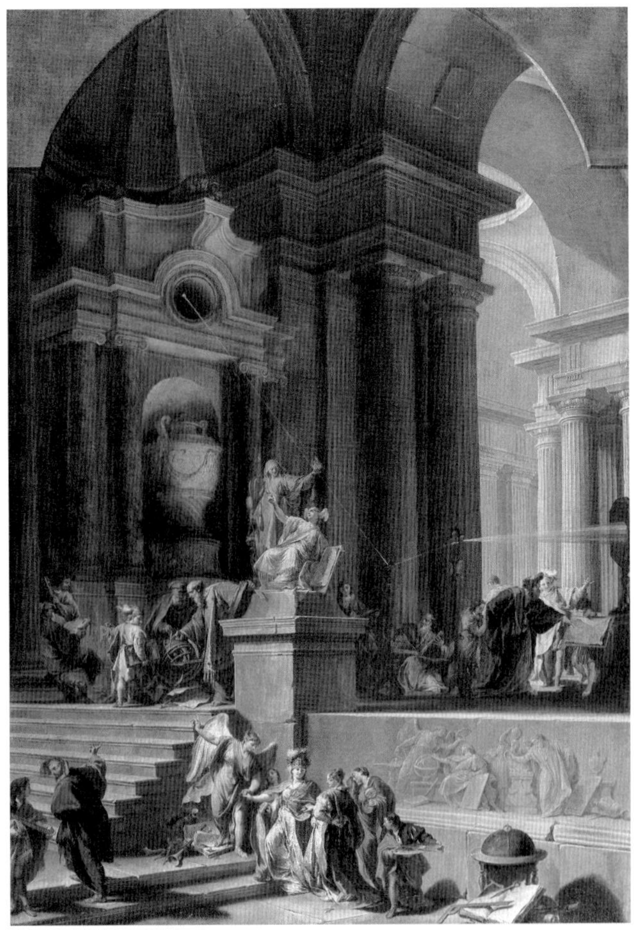

D.9
see colour Plate XXII

allegory of Newton's triumphs, with Minerva and the Sciences led weeping to the urn containing Sir Isaac's remains.[6] In the centre of the painting is a large pedestal on which the female figures of Mathematics and (possibly) Truth are placed, with alongside it various figures representing ancient and modern predecessors of Newton studying diagrams and instruments. A ray of light crosses the centre of the picture, which, after passing through a prism, has unfortunately been painted as bending, and Pittoni has not managed to paint the order of the resulting spectrum correctly. And see engravings **LI-1** and **LII-1**. The Fitzwilliam Museum, Cambridge, celebrated the acquisition in 1973 with an exhibition: 'The European Fame of Isaac Newton'.[6]

Giovanni Battista Pittoni (1687–1767)

D.10 *An Allegorical Monument to Sir Isaac Newton*, oil on canvas.
154 × 218 cm ($60\frac{5}{8} \times 85\frac{7}{8}$ in)
Marzotto Collection, Valdagno, Italy.

Provenance: Sale of the Collection M. F. B., Galerie Fiévez, Brussels, 14 May 1928 (lot 69) as 'Les Sciences et les Arts'; Sackville Gallery, London, 1928; Marzotto di Valdagno Collection. Published by Laura Coggiola in 1934: 'Altre opere inedite di Giambattista Pittoni' in *Rivista della Città di Venezia*.

This canvas, signed 'Pittoni Roma', is a variation of **D.9**, and follows suggestions for the design, both in format and content, in a rough sketch in John Conduitt's papers at King's College, Cambridge. Instead of the urn there is a representation of Newton's tomb in Westminster Abbey, with the surrounding, but simpler, architecture open to the sky. The picture also extends further to the right, where there is statuary in an apse. The two figures on the pedestal in the centre and the other groups of figures remain the same as in **D.9**, but at the lowest level there are more and different wall bas-reliefs.There is a new figure, which could be Aristotle,

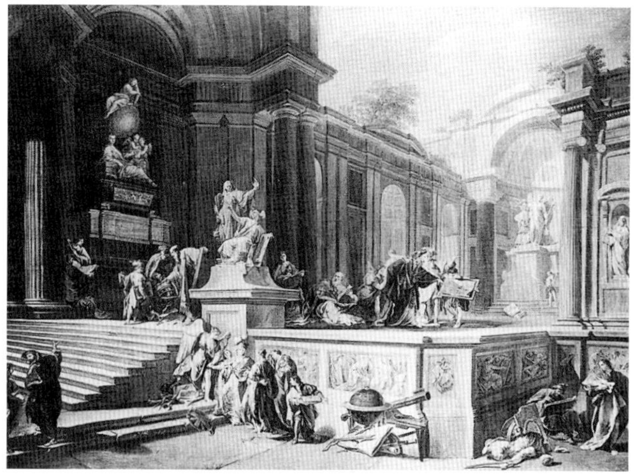

D.10

in a niche on the right above two figures, one seated working at a board on his knees and the other standing, in front of the bas-reliefs. There is no ray of light crossing the centre of the canvas.

Giacomo Ciesa from Vicenza

D.11 Fresco painted on the wall of the Upper Observatory, now La Specola Museum, of the Osservatorio Astronomico in Padua in 1772/3.

Life-size, full-length portraits of Ptolemy, Copernicus, Tycho, Galileo, Kepler, Newton, Montanari and Poleni are placed around the circular room, with above each, in smaller frames, monochrome chiaroscuro scenes and figures taken from Greek mythology. Newton holds in his right hand, between finger and thumb, a small sphere linked below to another larger one by a loop said to represent the impalpable, invisible ether through which gravity acts. His left hand rests on the top of the door frame, which cuts off part of his left side, on a copy of *Principia*. Above, a woman holding a lyre and representing Music sits on a globe, which is here said to signify celestial harmony. In discovering the universal physical cause which regulates the motion of celestial bodies, Newton had found the law of cosmic harmony. (See colour plate **XXIII**.)

Étienne Louis Boullée (1728–1799)

D.12 Design, never realised, for a cenotaph to Sir Isaac Newton. Six drawings in the Bibliothèque Nationale, Paris, one of which is dated 1784.

In his neo-classical design, Boullée planned to have a pure sphere encircled by two cylindrical rings. The tomb was to have a circular entrance, and be lit by daylight through apertures to give the appearance of a night sky from the light of the stars and planets. He wished to pay homage to Newton as the pre-eminent natural philosopher of his age, and thought of him as 'Esprit Sublime: Génie vast, et profond! Etre Divin! Newton! Daigne agréer l'hommage de mes faibles talents.' Elsewhere he wrote of Newton: 'You have determined the shape of the earth and I have conceived the project to enclose you within your own discovery', and invoked Newton as a hero for the opposite reasons from William Blake (**XXXI-I** and **XXXI-2**). See A. M. Vogt, *Boullée's Newton-Denkmal*.[38]

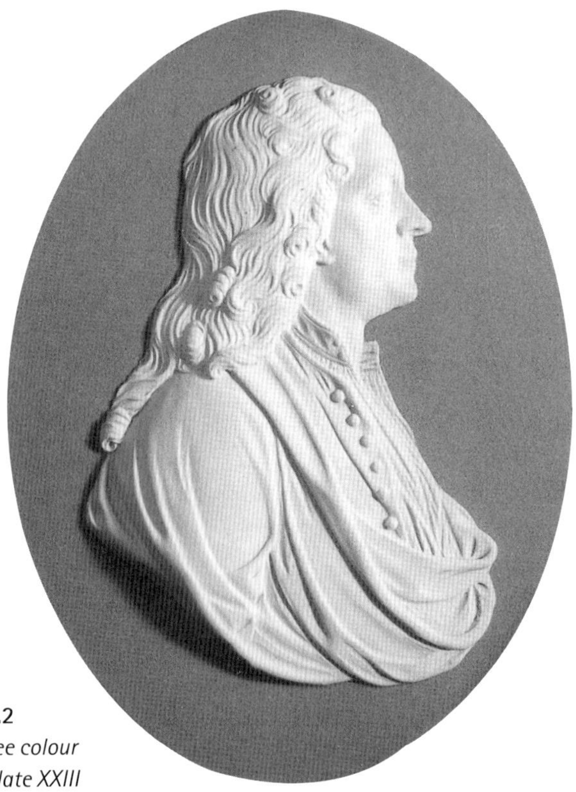

E.2
see colour Plate XXIII

E. CERAMICS AND WAXES

Wedgwood Portrait Medallions,[7,27] either in black basalt or in blue and white jasper, are listed in three sizes under 'Heads of Newton' in the section 'Philosophers, physicians &c' in the Wedgwood and Bentley Catalogue, 1779 [**Josiah Wedgwood (1730–1795)**; **Thomas Bentley (1730–1780)**]. It is of note that John Flaxman (1755–1826), who began to be employed by Wedgwood in 1775, presented a copy of Newton's death mask (**C.4**) to the Hunterian Museum, Glasgow University, in 1819.

E.1 Oval relief in black basalt with moulded basalt frame, 1773.
24.2 × 20 cm ($9\frac{1}{2} \times 7\frac{7}{8}$ in)

Unmarked, full-frontal bust, wearing long wig, after the ivory portrait plaque (**B.1**) by David le Marchand.
Lady Lever Art Gallery, National Museums and Galleries on Merseyside; Sotheby's 19 December 1963 (lot 165).

E.2 Oval relief, white on blue jasper.
11.1 × 8.5 cm ($4\frac{3}{8} \times 3\frac{3}{8}$ in)

Wigged profile bust facing to the right, wearing unbuttoned shirt and coat under loose outer garment. The mould at the Wedgwood Museum, Barlaston, is marked 'Newton'. The copy of the medallion in Spalding Gentlemen's Society is labelled as having been modelled by John Flaxman, who possessed a copy of Newton's death mask. The medallion in Horsham Museum has been missing since 1972.

E.3 Oval relief, white on blue jasper or black basalt.
9.2 × 7.6 cm ($3\frac{5}{8} \times 3$ in)

Unwigged profile bust looking to right, wearing unbuttoned shirt and coat under loose outer garment, with a comet, behind, at the level of the collar.
Spalding Gentlemen's Society; Tangye Collection, City Museum and Art Gallery, Birmingham.

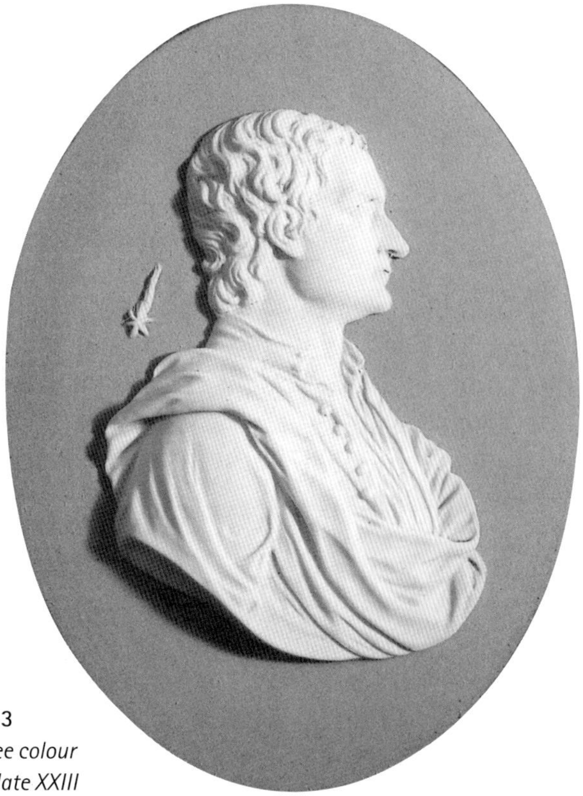

E.3
see colour Plate XXIII

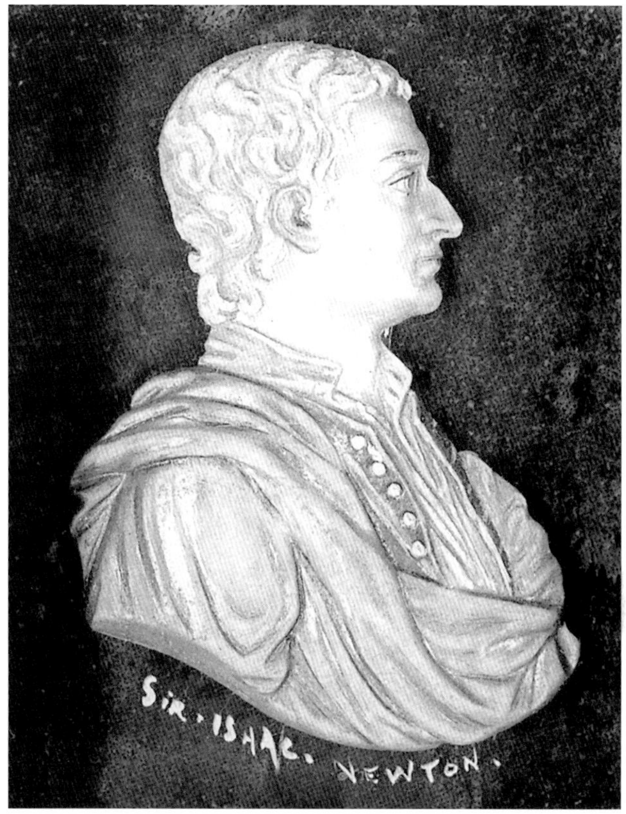

E.4

Possibly modelled by William Hackwood (*c.*1757–1839), but source unidentified.

E.4 Coloured wax cast, using same model from which **E.3** was made.
8.5 × 6 cm ($3\frac{3}{8} \times 2\frac{3}{8}$ in)
Wellcome Institute Library, London (no. 18352).

Plaster casts were made from a model fashioned in medal bronze, which could then be used to make a mould for a portrait medallion, or in this case for a wax cast.

E.5 Framed oval relief in reverse after **E.3**: unwigged, bust facing to the left, with comet behind head, in glazed white Continental creamware ? eighteenth century.
15 × 14 cm ($5\frac{7}{8} \times 5\frac{1}{2}$ in), 2.5 cm (1 in) deep
Henry Willett Collection, Royal Pavilion Art Gallery and Museums, Brighton since 1901.

The bust-relief has Newton wearing an unbuttoned shirt with a loose garment over it, within a wide frame.

E.6 Oval relief, 6 cm ($2\frac{3}{8}$ in), unwigged, profile of head and neck, with a comet placed behind the nape of the neck, white on blue jasper. Usually thought to be adapted from the medal (**G.4**) by Roëttiers of 1739, but a version has been reported marked 'W. Hackwood' on the truncation.
Haworth Collection, Manchester Museum, on loan to City Art Gallery, Manchester.

E.7 Oval relief, full-frontal bust, unwigged, head turned slightly to left, with comet at the right, white on blue jasper, or black basalt, dated 1779–80.
Height 27.5 cm ($10\frac{7}{8}$ in). The comet does not appear on all models. Mark: WEDGWOOD & BENTLEY.
British Museum (E.26 – 1887); Fogg Museum of Art, Cambridge, MA; Emily Winthrop Miles Collection, Brooklyn Museum, New York; Buten Museum, Sand Point Reserve, Port Washington, NY.

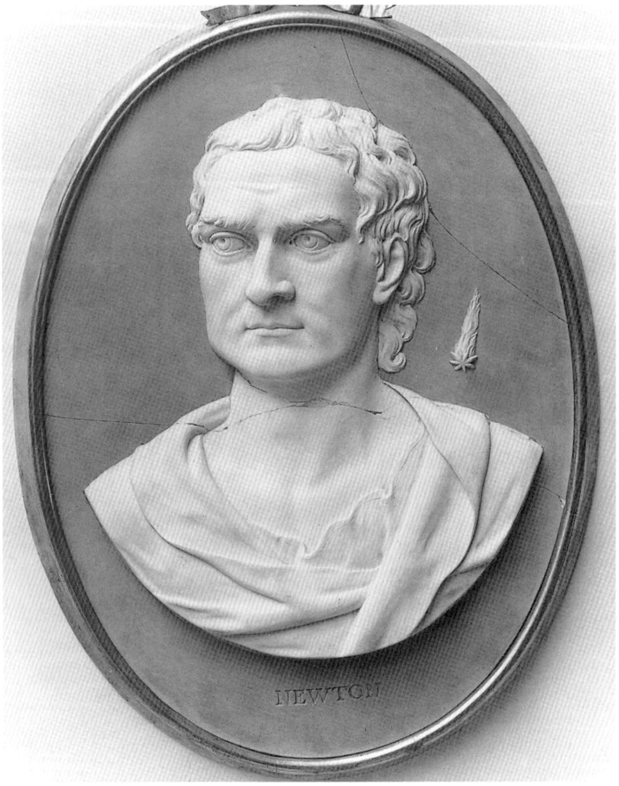

E.7
see colour Plate XXIII

This has often been thought to have been modelled after an untraced ivory carving by Silvanus Bevan (1691–1765), but can be securely attributed to William Hackwood, who made it after 'finishing' John Cheere's Type 2 plaster bust (**A.36**), which was invoiced to Wedgwood & Bentley in 1775 by Hoskins and Grant, and 'improving' it at the factory – see **E.8**.

Wedgwood and ceramic busts and figure

The making of ceramic busts, figures or reliefs involves casting and moulding. In his early years Wedgwood did not employ sculptors and relied on external sources for plaster models of a good likeness, which might be in several sizes; sometimes he borrowed from a private collection and moulded directly. A hollow mould, or matrix, would be made to reproduce the object in plaster, which would then be 'finished' in the factory workshop at Etruria, 'vastly improved', Wedgwood said, by William Hackwood and his other modellers, to make a patrix.

The patrix is then used to produce a series of working moulds, or matrices, which can be replaced when worn out by casting fresh moulds, maintaining standards of production, and allowing the original model to be preserved. The final product is smaller than the original model from shrinkage as the clay dries, or when it is fired. Advantage of the shrinkage can be taken by making further moulds to obtain reductions in size, a procedure that is comparatively easy and gives precise replicas in all but scale; enlargements require costly remodelling, and are unlikely to result in a replica of the same precision. Thus, while a 20 inch plaster might be used for the production of precise replicas in black basalt in smaller sizes, it could not be the model for a bust taller than 18 inches without being remodelled.

E.8 and **E.9** Two Wedgwood busts, on plain turned circular or shaped socles, in black basalt, **E.8** in 1774, and both **E.8** and **E.9** in 1777, 1779 and 1787 Wedgwood & Bentley catalogues.
In various sizes, 18–20 in, 14–15 in, 10–11½ in and 7–8 in high.
Wedgwood Museum; Emily Winthrop Miles Collection, Brooklyn Museum, New York; National Science Museum, London; Zeitlin Collection, Merion, Pennsylvania; Potteries Museum and Art Gallery, Stoke-on-Trent; Christie's 11 December 2000 (lot 59).

There were two plaster busts of Newton listed on an invoice, dated 1 January 1775 (see E. Meteyard, *The Life of Josiah Wedgwood*[39]), from James Hoskins (d.1791) and Benjamin Grant (*fl.*1775–1809) to Wedgwood & Bentley, which both originated with John Cheere and were after Roubiliac (**A.18**). It is not known after which John Cheere Types (2 or 3) – see (**A.36**) and (**A.38**) – these two busts of different but unknown sizes were that were sent to be finished at the Wedgwood factory, but one was supplied in time to allow the 18 inch bust **E.8** (Reilly, *Wedgwood*,[27] K31) to be listed in the 1774 catalogue. The bust illustrated (plate XL, fig. 2) in Grant, *The Makers of Black Basaltes*[27] was 17⅞ inches

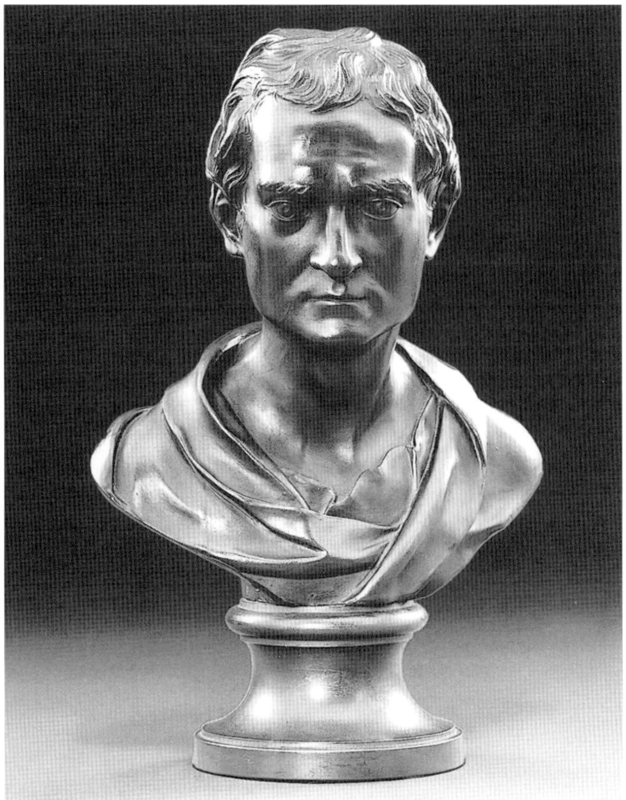

E.8

high and John Cheere Type 3 (**A.38**), and the one in Christie's 11 December 2000 (lot 59), 14 inches high, with 'NEWTON' impressed to reverse of bust and 'Wedgwood & Bentley' to both bust and socle, was finished from John Cheere Type 2 (**A.36**), as were **E.10** and **E.11**. Another reduced version, 15 inches high (Reilly K67), is listed in the 1777 catalogue.

E.9, the second Wedgwood bust of Newton (Reilly K72, 1777), is smaller than the first bust, **E.8**. It is 11 inches high with its socle (illustrated in Reilly, *Wedgwood*,[27] vol. I, p. 457, plate 653), and based after the ivory bust (with no socle) by David le Marchand (**A.2**) of 1718, which is $9\frac{3}{4}$ inches high and was given to the British Museum in 1765. It is unlikely that it was borrowed from the museum, and more likely that **E.9** was modelled or cast in a mould by one of the plaster makers such as John Cheere or Charles Harris – it was not supplied by Hoskins and Grant as it is not in any of their invoices which have survived pretty much intact – and required little finishing at Wedgwood. Reilly K79, 1777, is a reduced version of K72, and 7–8 inches high.

In the Wedgwood Museum, founded in 1906 principally from the informal collection at the Etruria factory, there are two black basalt busts, both **E.9**: black basalt bust no. 1104, 9 inches high, is marked 'WEDGWOOD'. It was bought Christie's 22 April 1885 (lot 87) for the Wedgwood factory, and went to the Wedgwood Museum on its establishment. Black basalt bust no. 5157 was purchased in 1950 for the Wedgwood Museum, and is also 9 inches high.

Two damaged plaster busts in the Wedgwood Museum came from the factory: plaster bust no. 4529 is **E.8**, 17 inches in height, but it is not known after which John Cheere Type (2 or 3) it was finished – see **E.10** and **E.11**. Plaster bust no. 4648 is a highly coloured **E.9**, 12 inches in height. It is not known why and when it was coloured.

E.10 Painted plaster bust on shaped socle, after John Cheere Type 2.
42.5 cm ($16\frac{3}{4}$ in) high including socle, 30.5 cm (12 in) wide
American Philosophical Society, Philadelphia (no. 58.S.16).

Scratch on bottom:

NEUTON

Brass tag on socle:

SIR ISAAC NEWTON / O. S. / 1642-1727

This bust is a larger version of **E.11**. Two plaster busts supplied by Hoskins & Grant, a London firm of plaster-cast makers, and invoiced 1 January 1775 to Wedgwood & Bentley – see E.8 – originated with John Cheere, for whom Hoskins and Grant had worked in 1751. Except for a different, rounded socle, **E.10** is very close to **E.8**, a bust also after John Cheere Type 2 (**A.36**). It appears to have been painted black before it was overpainted in white: there are four black-painted plaster busts illustrated in Reilly's *Wedgwood*,[27] vol. I, plate 655, and the shaped socle, taken from the Roubiliac bust, is of a type frequently used by Wedgwood & Bentley. It is recorded that many original models and casts were sold, or otherwise disposed of, from the Wedgwood factory by Josiah Wedgwood II (1769–1843) in October 1828, and this plaster bust might possibly be one of them.

E.11 Painted plaster bust on shaped socle, after John Cheere Type 2.
26.7 cm (10½ in) high
Department of Ceramics and Glass, Victoria and Albert Museum (no. A.10–1977).

Painted dirty white, and on middle of back inscribed:
Newton.

This is a smaller version of **E.10**, and, as suggested by Friedman and Clifford,[13] originated with John Cheere Type 2 (**A.36**), and is close to black basalt bust (**E.8**) finished in the Wedgwood factory. Its handling is similar in its style and detailing to Cheere's bust of Alexander Pope in the Victoria and Albert Museum. It, too, may have been painted black before being overpainted in white, and have come from the Wedgwood factory in 1828. With **E.10** it provided the model for the ceramic bust by Enoch Wood (**E.12**).

E.12 Ceramic bust, with waisted socle, by **Enoch Wood (1759–1840)**.
In two sizes: 30.5 cm (12 in) and 24 cm (9½ in) high
Victoria and Albert Museum (nos. 64 – 1874 and 88 – 1874); Henry Willett Collection, Royal Pavilion Art Gallery and Museums, Brighton, since 1901.

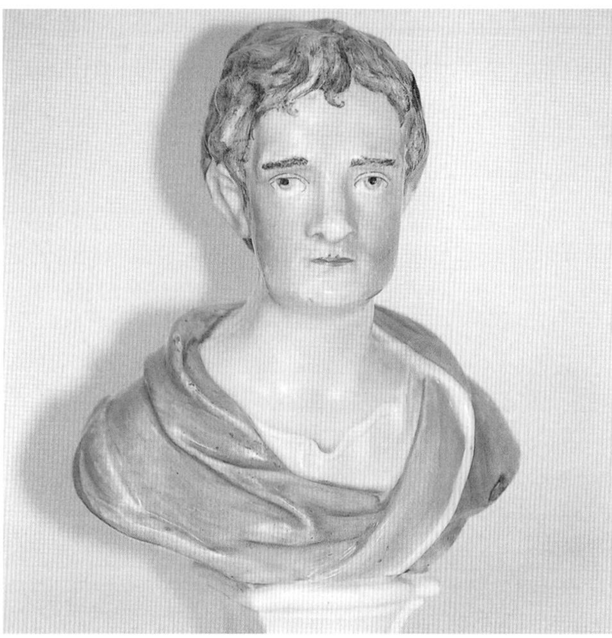

E.12
see colour Plate XXIV

Pearl glazed earthenware with overglaze enamel painted decoration, the shoulders being draped with emerald and cerise, and the socle painted to simulate marble. NEWTON impressed on back of socle.

The plaster busts (**E.10**) and (**E.11**), after John Cheere Type 2 (**A.36**), from Wedgwood may be thought to have provided the model for this coloured ceramic bust, which, though it is usually considered to have been done by Enoch Wood, might possibly be by Ralph Wood.

E.13 Fürstenberg porcelain bust, 1783[23,32]
16.1 cm (6¼ in) high
Hard-paste porcelain, the socle and pedestal glazed and gilt, marked NEWTON/No:1.
Fitzwilliam Museum, Cambridge.

E.14 Pottery figure by **Ralph Wood, the younger (1748–95)**, impressed 'Ra. Wood Burslem 137', or impressed 'LEEDS POTTERY', or by (Enoch) Wood and Cauldwell, Burslem, Stoke-on-Trent, using the same model, pearl glazed earthenware, plain white or with overglaze enamel painted decoration, *c.*1785–1800.[16,21]
26.7 cm (10½ in) high, or 31 cm (12¼ in) with socle.
Temple Newsam House, Leeds (no. 16.199/47); Fitzwilliam Museum, Cambridge; Trinity College, Cambridge; Grace K. Babson Collection on deposit in the Burndy Library, Dibner Institute, Massachusetts Institute of Technology; Henry Willett Collection in the Royal Pavilion Art Gallery and Museums, Brighton, since 1901; Potteries Museum and Art Gallery, Stoke-on-Trent.

Newton stands, turning to his right, on a high rectangular pedestal, with the name *Newton* on its rear, painted to imitate marble. He wears a pink dressing-gown lined with green, or white material with dark brown spots (ermine ?), over a sprigged waistcoat with blue borders, yellow breeches, white stockings and green, or red, shoes. On his left there is a celestial globe which rests on a pile of books. Between the leaves of one of the books is slipped a sheet of paper showing a comet

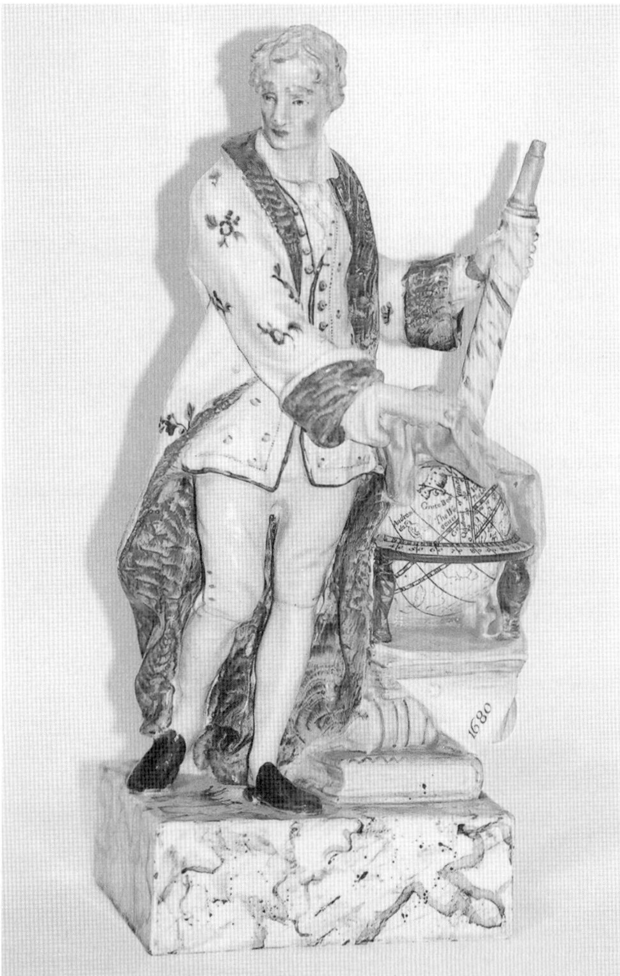

E.14
see colour Plate XXIV

and with the date 1680 on it. In his left hand he supports a telescope, and with his right he holds a brown drape over the globe. There is an apple against his left foot. The details in the modelling and colouring of the figure vary considerably in production.

Wax relief and images

Matthew (Mathieu) Gosset (1683–1744)

E.15 Oval, portrait relief in wax, bust-profile facing right, before 1729.
10.5 × 8.5 cm ($4\frac{1}{8} \times 3\frac{3}{8}$ in)
Spalding Gentlemen's Society.

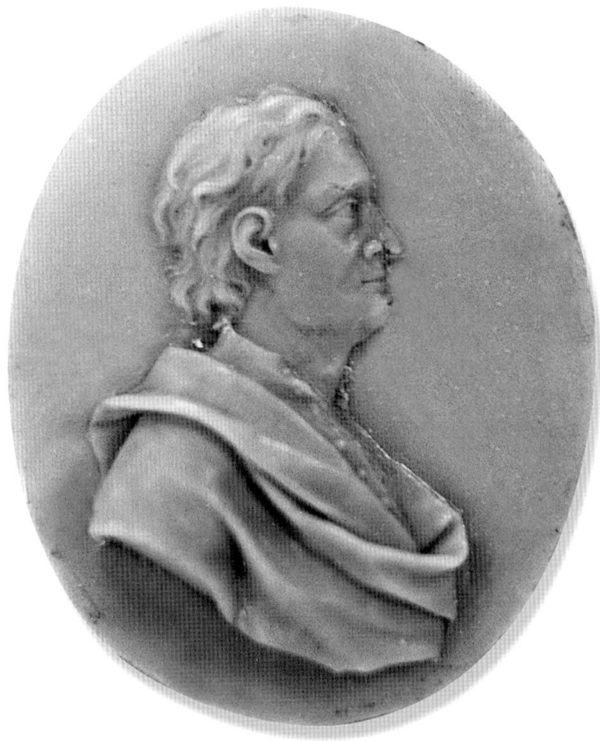

E.15

In the Minutes[24] of the Spalding Gentlemen's Society it is recorded that in 1729 George Johnson of Durham School presented to the Society's museum 'a curious effigy of the Hon. Sir Isaac Newton knight … made in profile in the manner of a medaglion by the ingenious Mr. Gosset.'

E.16 Unattributed, rectangular wax panel in high relief, late eighteenth century.
19 × 14.6 cm ($7\frac{1}{2} \times 5\frac{7}{10}$ in)
Schreiber Collection, Victoria and Albert Museum, London, since 1884;[25] Wellcome Institute Library, London.

Half-length; Newton is shown wearing an open-necked coat, loose robe and round cap; head half-turned to right, arms held out, the hands resting on a book at the base of a flat pilaster on the middle of which is an inscription of Alexander Pope's words, in high relief:

> Nature and Nature's Laws
> Lay hid in night.
> God said, Let Newton be
> And all was light.

E.17 Wax images of Newton mentioned in E. J. Pyke. *A Biographical Dictionary of Wax Modellers.*
Oxford: Clarendon Press, 1973, and its supplement.[24]

(a) **Isaac Gosset (the elder) (1713–1799)**
Portrait head (effigy modelled in wax) of Isaac Newton (1758), given by Thomas Hollis[18] in 1761 to Leiden University, and now in Academisch Historisch Museum der Universiteit Leiden.

(b) **Bernhard Caspar Hardy (1726–1819)**
Wax relief of Isaac Newton in Esztergom: Keresztény Múzeum.

(c) **Samuel Percy (1750–1820)**
Wax relief of Isaac Newton with Prof. Mario Praz in Rome.

(d) **John Sigismund Tanner (*c.*1706–1775)**
A small head of Isaac Newton modelled in wax, now missing. John Conduitt introduced Tanner to work at the Mint.

F. CAMEOS AND INTAGLIOS[22]

F.1 Wedgwood cameos, 'either in a fine *Waxen biscuit* of extreme Hardness, with Grounds enameled of different colours; or in the *artificial Jasper* with blue or brown Grounds ... Or they may be had in the white waxen Biscuit, or in blue Jasper.'
Sir Isaac Newton's head, nos. 1685, 1686 and 1687, first appeared in 1777 Wedgwood & Bentley catalogue.

F.2 Wedgwood intaglios in black basaltes: Sir Isaac Newton's head, nos. 20, 28 and 96, first appeared in 1774 Wedgwood & Bentley catalogue. No. 96 is described as 'Sir Isaac Newton, from his own ring', and is smaller.
Grace K. Babson Collection on deposit in Burndy Library, Dibner Institute, Massachusetts Institute of Technology.

F.3 Newton with comet, facing quarter-left, set in gold ring made *c.*1790. The stone measures 2.1 × 1.6 cm.
Grace K. Babson Collection on deposit in Burndy Library, Dibner Institute, Massachusetts Institute of Technology.

Around edge of ring:
Presented to Sir John Herschel [1792–1871].

F.4 Cameo portrait of Newton in onyx, of upright oval shape, mounted in gold with a suspension ring above and a bracket below.
2 × 1.75 cm, with overall size of the object (ring open): 36 × 18.5 cm
Devonshire Collection, Chatsworth, Derbyshire.

Christopher Seaton, gem engraver, was paid 25 guineas for a head of Newton exhibited at the Society of Artists in 1768.[17,30]

F.5 Intaglio bust of Newton engraved on one of three faces of triangular rock crystal mounted in gold, scrolled swivel-fob, eighteenth century.
2.1 × 1.7 cm
Fitzwilliam Museum, Cambridge (PER 156 (M)) since 1922.

Newton is bareheaded and in contemporary dress, facing in profile to the left, with an ascending comet placed behind the nape of his neck. The initials REH in monogram on second face.

Anthony Shepherd, FRS (1721–96), Plumian Professor of Astronomy at Cambridge University, presented an intaglio portrait of Newton set in a ring to the Astronomer Royal, Nevil Maskelyne, FRS (1732–1811), which was exhibited by his grandson, M. H. N. Story-Maskelyne, FRS (1823–1911) at the Archeological Institute in 1861.[17]

F.6 Sixteen cameo heads and busts of Newton, nos. 14312–14328 and 15780, in the enamel paste of James and William Tassie, from gems engraved by **James Tassie (1735–1799)**, **Robert Bateman Wray (1715–1779)** and **Edward Burch (1730–1814)** are listed in R. E. Raspe's Catalogue of 1791.[26]

Medallion no. 14312: Cameo head in enamel paste, 3.5 cm, marked *Tassie Glasguensis, F.*, 'done in the antique manner; face with prominent drooping nose, to right; hair in curling locks; ear prominent'. The description continues: 'A head of Newton ... belongs to Mr John Anderson, Professor of Natural Philosophy in the University of Glasgow, and with a purse of gold, is annually given by him, for the encouragement of Natural Philosophy in that Society. The Paste is fixed in an elegant frame of silver, with ... inscription upon the margin.' One of two medallions given annually as a prize in the University of Glasgow for the best Physical Essay.

F.7 Intaglio bust of Newton, facing left, with long hair, within three oval lines. Cut in paste by **James Tassie**.
2.4 × 2 cm
Grace K. Babson Collection on deposit in Burndy Library, Dibner Institute, Massachusetts Institute of Technology. Formerly belonged to wife of Sir William Huggins (1824–1884).

F.8 Wax impression of oval seal depicting Sir Isaac Newton cut by **James Tassie**.
1.9 × 1.8 cm
Wren Library, Trinity College, Cambridge (King C6, 39(2)).

G. PORTRAIT MEDALS AND TOKENS

There are four posthumous portrait medals of Newton showing his image[33], one by J. Croker in 1726 (Old Style), two by Jean Dassier (not dated), and one by J. Roëttiers struck in 1739, which was copied later in 1774.[34] There are specimens in British Museum, Fitzwilliam Museum, Trinity College and King's College, Cambridge.

G.1 Portrait medal by John Croker (1670-1741), 1731

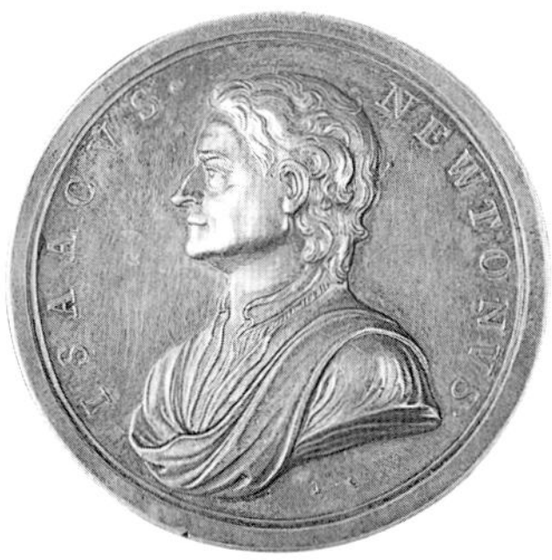

Obverse: Profile bust of Newton facing left, hair short, in shirt with open collar and mantle round the shoulders.
Legend:
ISAACUS . NEWTONUS
with below,
I. C.

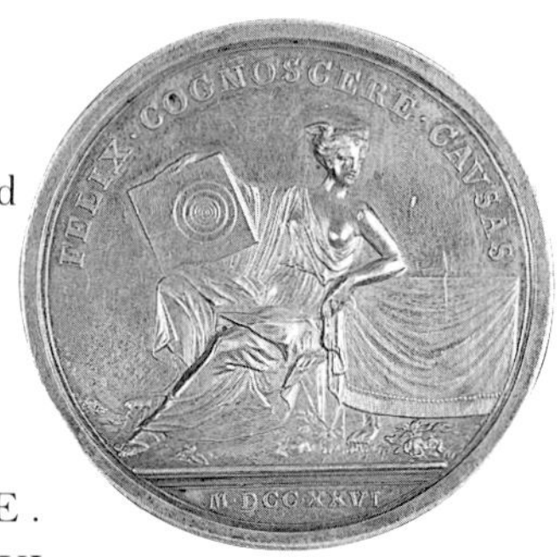

Reverse: Science, with wings on her head, seated facing left, leans upon a table and holds a plan of the solar system.
Legend:
FELIX . COGNOSCERE . CAVSAS / M. DCC. XXVI.
(legend abridged from Virgil, *Georgica* II, line 490: 'Felix qui potuit rerum cognoscere causas'; 'Happy he who could know the causes of things').

5.1 cm in diameter, in silver, copper or bronze. The legend alludes to Newton's sagacity in penetrating the primary causes which governed the motion of the planets, and the device to his elucidation of the solar system. The medal was struck in 1731. John Croker was Chief Engraver at the Mint from 1705.

G.2 Portrait medal by Jean Dassier (1676–1763)

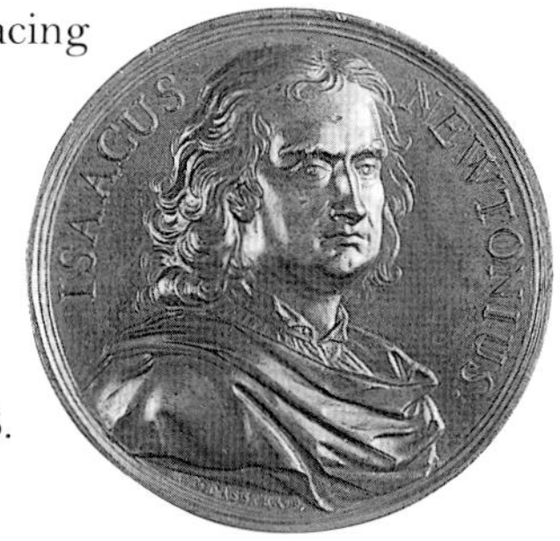

Obverse: Bust of Newton facing right, hair long, in shirt with open collar, and mantle round the shoulders.
Legend:
ISAACUS NEWTONIUS.
with below,
I. DASSIER. F.

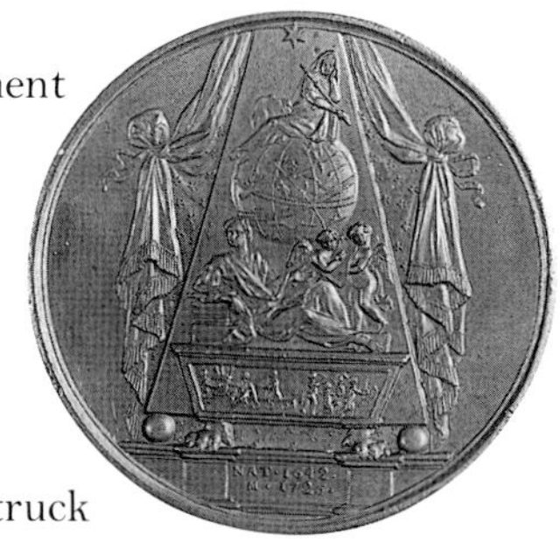

Reverse: Newton's monument in Westminster Abbey, with the base inscribed,
NAT. 1642 . M . 1726.

4.2 cm in diameter, in silver, bronze or gilded copper. This medal was struck after the unveiling of the monument in April 1731, with Newton's death being given as 1726.

G.3 Portrait medal by Jean Dassier (1676–1763)

Obverse: Bust of Newton looking right, hair long, in shirt with open collar and mantle round the shoulders.
Legend:

ISAACUS NEWTONIUS

Reverse: A wreath of flowers enclosing the inscription
EQ. AUR. / PHILOSOPHUS. / OBIIT 31. MART. / 1727. / NATUS ANNOS / 85.

3.2 cm in diameter, in silver or copper. The medal appears to have been struck after 1752 as the date of death is given after the New Style.

G.4 Portrait medal by Jacques Roëttiers (1707–1784), 1739

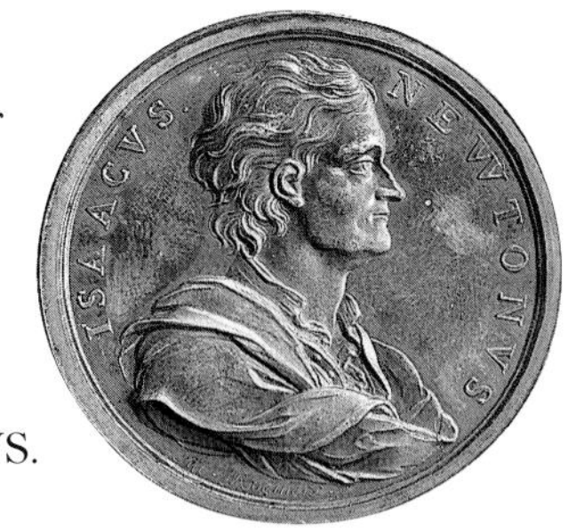

Obverse: Profile bust of Newton, facing right, hair short, in shirt with open collar and mantle round the shoulders.
Legend:
ISAACUVS. NEWTONVS. with below,
Iac. Roettiers.

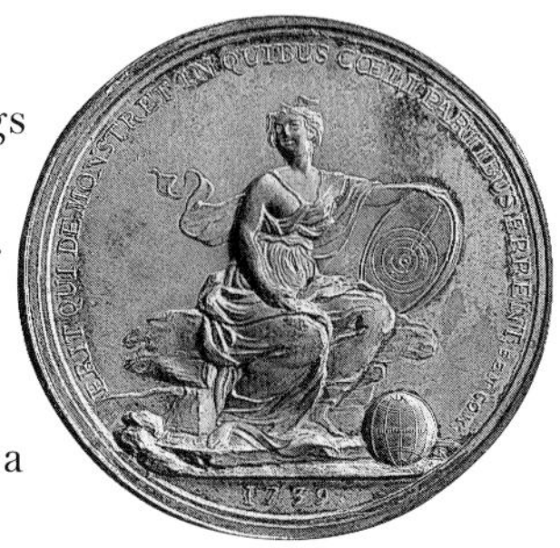

Reverse: Science, with wings on her head, seated facing right upon a rock with her head turned to left and holding a diagram of the solar system; at her feet is a globe.
Legend:
ERIT QUI DEMONSTRET IN QUIBUS COELI PARTIBUS ERRENT. SEN. COM: / 1739. R.
('There will be someone to demonstrate in what part of the heavens they wander' – from Seneca's *De Cometis.*).

5.3 cm in diameter, in silver or bronze. This medal was executed by order and at the expense of Thomas Hammond, and only a few were struck off as presents to his friends and to admirers of Newton. John Conduitt lent Roëttiers drawings of Newton on which to base the image on this posthumous medal. The device of the reverse is very similar to that on Croker's medal, and the legend considers Newton as fulfilling the prophecy of Seneca (*Quaestiones Naturales.* Liber vii. Cap. 26) that someone would arise to demonstrate the course of the comets.

G.5 Medal after Jacques Roëttiers, 1774, medallist unknown

Obverse: Draped bust of Newton, short hair, facing right.
Legend:
ISAACUS NEWTONIUS, with below, 'Jac. Roëttiers.

Reverse: QUAERITUR / HUIC / ALIUS / VIRG . AENEID : / M. DCC. LXXIV
('Another is sought [in vain] in his place' – from Virgil's *Aeneid*).

5.4 cm. in diameter in bronze or silver. This piece is a class prize for Natural Philosophy at Glasgow University. The obverse design is taken from the medal (**G.4**) by J. Roëttiers struck in 1739.

G.6 Portrait medal, anonymous, 1767

Obverse: Bust right. ISAACUS NEWTONUS; in field below bust, VERONAE MDCC/LXVII

Reverse: An eagle soaring towards the sun in splendour over an aerial view of the British Isles. Around: VOLUCRES PRAETERVOLAT OMNES
(He flies beyond all birds).

Cast and chased, 11.5 cm in diameter. This medal was probably made for an academic occasion in Verona in 1767.[6]

The tokens[35] are either halfpennies or farthings, made of bronze or copper, and of variable quality, dated 1793. Specimens are in British Museum, Fitzwilliam Museum, Trinity College, Cambridge, National Portrait Gallery, National Museum of Science, and Grantham Museum.

G.7 Halfpennies

Obverse: Bust of Newton, facing left.
Legend:

S^{R} ISAAC NEWTON

Reverse: A caduceus, olive branch and cornucopia.
Legend: HALFPENNY. 1793.

The right-hand wing of caduceus nearly touches the second limb of the first N of 'HALFPENNY'.

The edge of the token is roughened or smooth, with, on some, 'Payable in London Bristol & Lancaster.', or 'Payable at the Warehouse Liverpool.' 2.8 cm in diameter and 2 mm thick.

G.8 Farthings

Obverse: Bust of Newton facing to left.
Legend:

S^{R} ISAAC NEWTON.

Reverse: A cornucopia and olive branch.
Legend:

FARTHING. 1793.

Edge may be milled or plain, 2.1 cm in diameter, 2 mm thick.

One has on the reverse a female seated holding spear and palm branch, and the date 1771, another has a shield bearing the Prince of Wales's plume, between sprigs, with legend PRO BONO PUBLICO 1795., others a variation in the bust of Newton.

BIBLIOGRAPHY

1. Charles Avery. *David le Marchand 1674–1726: An Ingenious Man for Carving in Ivory*. London: Lund Humphries Publishers, 1996.
2. *A Descriptive Catalogue of the Grace K. Babson Collection of the Works of Sir Isaac Newton and the Material Relating to Him in the Babson Institute Library, Babson Park, Massachusetts.* With an introduction by Roger Babson Webber. Published by Herbert Reichner, New York, 1950; *A Supplement to the Catalogue of the Grace K. Babson Collection of the Works of Sir Isaac Newton and Related Material in the Babson Institute Library, Babson Park, Massachusetts.* Compiled by Henry P. Macomber. Published by Babson Institute, 1955.
3. Malcolm Baker. 'The making of portrait busts in the mid-eighteenth century: Roubiliac, Scheemakers and Trinity College, Dublin'. *Burlington Magazine* 1995; 137: 821–31; Ingrid Roscoe. 'Peter Scheemakers'. *The Walpole Society* 1999; vol. 61: pp. 163–304, pp. 267–8.
4. Malcolm Baker. 'The portrait sculpture' in *The Making of the Wren Library, Trinity College, Cambridge.* Edited by David McKitterick. Cambridge: Cambridge University Press, 1995, p. 125 and notes 62, 63.
5. D. Bindman and M. Baker. *Roubiliac and the Eighteenth-Century Monument.* New Haven and London: Paul Mellon Centre for Studies in British Art / Yale University Press, 1995.
6. Catalogue of an exhibition *The European Fame of Isaac Newton* in the Fitzwilliam Museum, Cambridge, 22 November 1973 – 6 January 1974; Francis Haskell. 'The apotheosis of Newton in art' in The Annus Mirabilis of Sir Isaac Newton Tricentennial Celebration. *The Texas Quarterly*, Autumn 1967, 10, no. 3, pp. 218–37; and in F. Haskell. *Past and Present in Art and Taste: Selected Essays.* New Haven and London: Yale University Press, 1987.
7. A. Dawson. *Masterpieces of Wedgwood.* London: British Museum Press, 1984, p. 82, fig. 65.
8. A. Dawson. *Portrait Sculpture: A Catalogue of the British Museum Collection 1675-1975.* London: British Museum Press, 1999, no. 57, pp. 150–2; no. 58, pp. 152–4, and fig. 42; no. 59, pp.154–6, and colour plate 2.
9. K. A. Esdaile. *Roubiliac's Work at Trinity College Cambridge.* Cambridge: Cambridge University Press, 1924, pp. 2–5, 40–1.
10. K. A. Esdaile.*The Life and Works of Louis François Roubiliac.* Oxford: Oxford University Press, 1928, pp. vii, 19 note 1, 20, 21 note 1, 100–3, 106 note 1, 170–1, 214 note 1; E. J. Climenson. *Elizabeth Montagu – The Queen of the Blue-Stockings. Her Correspondence from 1720 to 1761.* 2 vols. London: John Murray, 1906, vol. II, pp. 35–6; J. V. G. Mallet. 'Some portrait medallions by Roubiliac'. *Burlington Magazine* 1962; 104: 153–8.
11. K. A. Esdaile. *The Art of John Michael Rysbrack in Terracotta*, illustrated catalogue. London: Spink & Son, 1932, pp. 35–6; J. A. Rouquet. *État des Arts en Angleterre / The Present State of the Arts in England,* Paris and London, 1755. Section: 'On sculpture' in facsimile reprint, with introduction by R. W. Lightbown. London: Cornmarket Press, 1970, pp. 62–5; Ernst Benkard. *Das Ewige Antlitz.* 1st edn, Berlin: Frankfurter Verlags-Anstalt, 1927;*Undying Faces: A Collection of Death Masks.* 2nd edn trans. by M. M. Green. London: Hogarth Press, 1929, with Note by Georg Kolbe: 'How death masks are taken', pp. 43-5, 72-3, plate 15; L. Hutton. *Portraits in Plaster, from the Collection of Laurence Hutton,* New York, 1894.
12. K. Eustace. *Michael Rysbrack: Sculptor 1694–1770.* Catalogue of an Exhibition in the City of Bristol Museum and Art Gallery, 6 March – 1 May 1982; 'The politics of the past: Stowe and the developments of the historical portrait bust.' *Apollo,* July 1998: 31–40.
13. T. Friedman and T. Clifford. *The Man at Hyde Park Corner, Sculpture by John Cheere 1700-1787.* Catalogue of exhibition at Leeds, Temple Newsam,

15 May – 15 June, and Twickenham, Marble Hill House, 19 July – 8 September 1974, introduction and nos. 21, 74, 97, pl. 26; M. I. Webb. 'Henry Cheere, sculptor and businessman and John Cheere – I and II'. *Burlington Magazine* 1958; 100: 232–40, 274–9; Charles Harris. *A Catalogue of the Statues, Bass Reliefs, Bustos, etc, of Charles Harris, Statuary, Opposite to the New Church in the Strand.* London, *c.*1790.

14. Christiano Giometti. 'Giovanni Baptista Guelfi: new discoveries'. *The Sculpture Journal* 1999; 3: 26–43; H. M. Colvin and J. Newman. 'Part II. The Royal Palaces: Richmond Lodge and Kew House' in H. M. Colvin (ed.). *The History of the King's Works.* V, 1660–1782. London: HM Stationery Office, 1976, pp. 217–29, pp. 224, note 4.
15. Derek Gjertsen. *The Newton Handbook.* London: Routledge and Kegan Paul, 1986.
16. P. Halfpenny. *English Earthenware Figures 1740–1840.* Woodbridge: Antique Collectors' Club, 1991, pp. 140, 173.
17. M. Henig, D. Scarisbrick and M. Whiting. *Classical Gems, Ancient and Modern Intaglios and Cameos in the Fitzwilliam Museum, Cambridge.* Cambridge: Cambridge University Press, 1994, p. 308, no. 644.
18. *Memoirs of Thomas Hollis Esq.* 2 vols. London: privately printed, 1780, vol. I, pp. 490–1.
19. A. Jacobs. *Laurent Delvaux – Gand, 1696 – Nivelles, 1778.* Paris: Arthena, 1999, pp. 238–40, 496–7, 507, 509.
20. R. Lane Poole. *Catalogue of Portraits in the Possession of The University, Colleges, City, and County of Oxford compiled by Mrs. R. Lane Poole.* 3 vols. Oxford: Clarendon Press, vol. I, 1912 and vols. II and III, 1926, vol. I, p. 91.
21. G. Lewis. *A Collector's History of English Pottery.* 3rd edn. Woodbridge: Antiques Collectors' Club, 1985.
22. Thomas Maude. *Viator, A Poem: or, A Journey from London to Scarborough, By the Way of York. With Notes Historical and Topographical.* London: B. White, T. Becket and J. Walter, 1782, pp. viii–xi.
23. J. E. Poole. *Plagiarism Personified?* Cambridge: Cambridge University Press, 1986.
24. D. M. Owen. 'The Minute-Books of The Gentlemen's Society 1712–55'. *The Lincoln Record Society* 1981; 73: xv; E. J. Pyke. *A Biographical Dictionary of Wax Modellers.* Oxford: Clarendon Press, 1973, pp. 56–7, 63–5, 103, 145–6, 202, and its supplement, 1981, pp. 31–2.
25. B. Rackham. *Catalogue of English Porcelain, Earthenware, Enamels, and Glass Collected by Charles Schreiber Esq., M.P. and The Lady Charlotte Schreiber and Presented to the Victoria and Albert Museum in 1884,* vol. II *Earthenware,* p. 124, no. 641, ill. plate 82.
26. R. E. Raspe. *A Descriptive Catalogue of a General Collection of Ancient and Modern Engraved Gems, Cameos as well as Intaglios, taken from the most celebrated Cabinets in Europe; and cast in Coloured Pastes, White Enamel, and Sulphur, by James Tassie, Modeller; arranged and described by R. E. Raspe; and illustrated with Copper-plates.* 2 vols. London. Printed for and sold by James Tassie and J. Murray, 1791; J. M. Gray. *James and William Tassie. A Biographical & Critical Sketch, with a Catalogue of Their Portrait Medallions of Modern Personages.* Edinburgh: W. Greenoak Patterson, 1894; J. P. Smith. *James Tassie 1735–1799: Modeller in Glass.* London: Mallett, 1995, p. 14.
27. Robin Reilly and George Savage. *Wedgwood: The Portrait Medallions.* London: Barrie and Jenkins, 1973, pp. 257–9; Robin Reilly. *Wedgwood Portrait Medallions.* An introduction to an exhibition at the National Portrait Gallery, 3 October 1973 to 5 January 1974. London: Barrie and Jenkins, 1973, no. 75; Wolf Mankowitz. 'A transcript of the 1779 Wedgwood and Bentley Catalogue' in *Wedgwood.* 3rd edn. London: Barrie and Jenkins, 1980, pp. 197–252; Robin Reilly. *Wedgwood.* 2 vols. London: Macmillan, 1989, vol. I, pp. 452–3, 454, 457, plates 644, 653, 655; vol. II, Appendix K; M. H. Grant. *The Makers of Black Basaltes.* Edinburgh: William Blackwood & Sons, 1910, p. 169, plate XL, fig. 2.
28. George Vertue. *The Notebooks of George Vertue,* I–VI. *The Walpole Society.* Vols. 18, 20, 22, 24, 26, 30; index

of I–V in 29, 30. Oxford, 1930–55. Vertue *Notebooks*, II: 20, pp. 69–70; III: 22, pp. 50–1, 57, 66, 71, 73–4, 133, 135; IV: 24, pp. 69–70, 166.

29. F. M. Arouet de Voltaire. *Lettres Philosophiques. Quinzième Lettre: Sur le Sistème de l'Attraction.* Gustave Lanson (ed.) 5th edn. 2 vols. Paris: Libraire E. Droz, 1937, pp. 15–29 (p. 24).
30. Horace Walpole. *Anecdotes of Painting in England: with some account of the principle artists; and incidental notes on other arts; Collected by the late Mr George Vertue And now digested and published from his original MSS. By Mr Horace Walpole.* 4 vols. Printed by Thomas Farmer at Strawberry-Hill. 1762–71, vol. IV, pp. 98–9; A. Graves. *The Society of Artists of Great Britain 1760–1791* and *The Free Society of Artists 1761–1783.* London: George Bell & Sons, 1907; R. Gunnis. *Dictionary of British Sculptors 1660–1851.* New revised edition. London: The Abbey Library, 1968, I. Gosset, M. Gosset, S. Percy, A. van der Hagen.
31. M. I. Webb. 'Busts of Sir Isaac Newton'. *Country Life*, 25 January 1952, pp. 216–18; *Michael Rysbrack – Sculptor.* London: Country Life Ltd, 1954; Appendix One: Letters I and IX from Michael Rysbrack to Sir Edward Littleton, pp. 194, 199; 'Giovan Battista Guelfi: an Italian sculptor working in England'. *Burlington Magazine* 1955; 92:139–45; M. Whinney. *Victoria and Albert Museum: English Sculpture 1720–1830.* London: HM Stationery Office, 1971, no. 5, p. 38.
32. B. F. R. F. Wolff Metternich. 'Die Porträtbüsten der Manufaktur Fürstenberg unter dem Einfluss der Kunstkritik Lessings', *Keramos* 1981: 19-68, at pp. 24, 44.
33. E. Hawkins. *Medallic Illustrations of the History of Great Britain and Ireland to the Death of George II.* Ed. A. W. Franks and H. A. Grueber. 2 vols. London: British Museum, 1885; illustrated editions 1904, 1911; reprinted by Spink & Son, 1969, vol. II, pp. 469–73 ; reprinted, London: Seaby Publications, 1979, plate CXLVII.
34. L. Brown. *A Catalogue of British Historical Medals 1760–1960.* Vol. I: *The Accession of George III to the Death of William IV.* London: Seaby Publications, 1980, p. 45.
35. J. Atkins. *The Tradesmen's Tokens of the Eighteenth Century.* London: W. S. Lincoln & Son, 1892, pp. 372–3, 382.
36. William Stukeley. *Memoirs of Sir Isaac Newton's Life.* Edited by A. Hastings White. London: Taylor and Francis, 1936, p. 85; J. Edleston (ed.) *Correspondence of Sir Isaac Newton and Professor Cotes.* London: J. W. Parker and Cambridge: J. Deighton, 1850, p. lxxxi.
37. Oliver Goldsmith. *The Life of Richard Nash, of Bath, Esq.* 2nd edn. London and Bath: J. Newbery and W. Frederick, 1762.
38. A. M. Vogt. Boullée's *Newton-Denkmal.* Basle: Sakvalbau und Kugelidee, 1969.
39. E. Meteyard. *The Life of Josiah Wedgwood.* 2 vols. London: Hurst and Blackett, 1865-6, vol. II, pp. 325-6.

INDEX

Note: The names of the owners of the icons are not included in this index

PHOTOGRAPHIC CREDITS

The author is grateful to all museums, galleries, other institutions and private owners who kindly supplied or allowed illustrations for this book. Where further acknowledgement is due is listed below with the sources for the photographs.

Arti Doria Pamphilj, Rome: **XIX-1**. Bonham's: **XI-2**, **XI-3**. ©British Museum: **V-1**; **VIII-3**; **XVII-1**; **A.2**; **A.16**; **B.11**; **D.3**; **E.7**; **G.4**. Michael Chevis: **IX-1**. Christie's: **XIII-4**; **A.33**; **E.8**. ©Courtauld Institute of Art Photographic Survey: **XXX-4**; **A.36**. Roy Davids: **A.6**. Dan Duchars: **A.34**. Richard Farleigh: **XXIV-3**. ©Fitzwilliam Museum, Cambridge: **IV-6**; **XXIV-1**; **XXXI-1**; **A.18**; **A.21**; **D.9**; **D.10**; **G.1**; **G.3**. ©UK Government Art Collection: **XXII-1**. A. Rupert Hall **X-1**. A. Rupert Hall **X-1**. Cecil Higgins Art Gallery, Bedford: **A.15**. ©Ilchester Estates: **A.11**. Milo Keynes: **XX-1**; **A.12b**; **A.22**; **A.31**. Simon Keynes: **A.12a**. ©King's College, Cambridge: **B.4**; **D.2**. The Library Company of Philadelphia: **XXXIV-1**. Louth Naturalists', Antiquarian and Literary Society: **A.30**. Earl of Mansfield: **A.8**. Ministero dei Beni e la Attività Culturali di Ferenze: **III-7**. ©National Portrait Gallery Photo Library: **Figure 4**; **III-1**; **XIV-4**; **A.3**. ©National Trust Picture Library: **XXIII-1**. Chris Rice: **A.38**. Royal Astronomical Society: **A.25**. Royal Collection ©2004 HM Queen Elizabeth II: **II-3**; **XVI-2**; **A.9**. Royal Pavilion Art Gallery, Brighton: **E.12**; **E.14**. ©Royal Society, London: **VI-1**; **XI-1**; **XIII-1**; **XIII-2**; **XV-1**. Spalding Gentlemen's Society: **B.19**; **E.2**; **E.3**; **E.15**. Late Alfred Spero: **A.1**. Richard Stanton: **XIV-2**. ©Tate, London 2004: **XXXI-2**. Lord Thomson of Fleet: **B.1**. ©Trinity College, Cambridge: **Figure 1**; **Figure 2**; **IV-1**; **IV-5**; **VII-1**; **XIII-3**; **XIV-1**; **XIV-3**; **XIX-1**; **XXIV-4**; **LVIII-1**; **Plate XII**; **A.14**; **A.24**; **C.3**; **D.7**. ©V & A Picture Library: **D.4**. ©Wellcome Library, London: **XXXVIII-1**; **XLIII-1**; **XLIV-1**; **XLV-1**; **XLVII-1**; **XLIX-1**; **LIV-1**; **LV-1**; **D.6**; **E.4**. ©Dean and Chapter of Westminster: **D.5a**, **b**, **c**, **d**. Jeremy Whitaker: **II-1**; **IV-2**; **XXXII-1**; **A.10**. White House Historical Association, Washington DC: **A.17**. James P. Williams: **VIII-1**. Yale Center for British Art, Paul Mellon Collection: **XXI-1**.